Fruits
of the
Desert

Prickly pear fruit

On the cover

The handsome foods on the cover, made with a variety of fruits that grow in the Sonoran Desert, are set off by Papago, Pima and Maricopa Indian baskets and pottery. Dishes and the pages where their recipes can be found are: apricot-glazed fruit tarts, Page 85; festive prickly pear jellied salad, Page 27; and sour orange bread, Page 68. In the jars are calamondin marmalade, Page 47 (standard method) or Page 171 (microwave); pomegranate jelly, Page 134; and grapefruit-sour orange marmalade, Page 44.

Library of Congress Cataloging in Publication Data

English, Sandal.
  Fruits of the Desert.
  Includes bibliography and index.
  1. Cookery (Fruit) 2. Cookery (Nuts) 3. Fruit — Sonoran Desert. 4. Nuts — Sonoran Desert. I. Title.
TX811.E53   641.6'4   81-20588   AACR2

ISBN 0-9607758-0-3

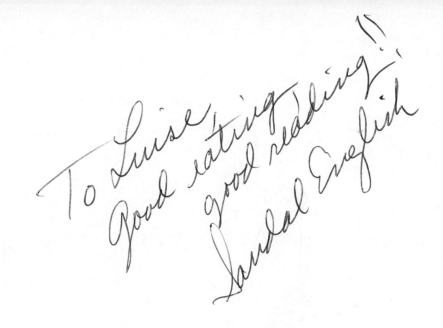

*To Luise
Good eating!
good reading!
Sandal English*

Fruits of the Desert

By Sandal English, food editor

The Arizona Daily Star

Star production staff

Layout and illustrations, Judy Margolis
Editing, Susan Albright, John Peck,
Debbie Kornmiller and Star copy desk
Composition, David Skog
Color photography, Scott Braucher

Published by The Arizona Daily Star
Tucson, Arizona

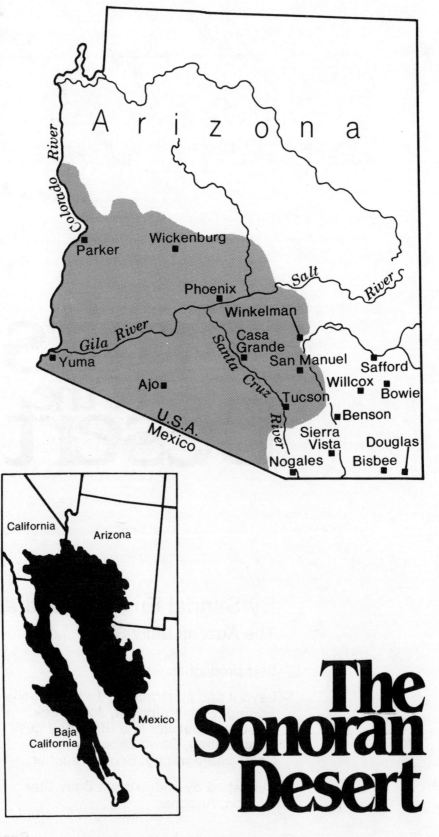

Arizona

Colorado River

Parker ■

Wickenburg ■

Phoenix ■

Salt River

Winkelman ■

Gila River

Casa Grande ■

Santa Cruz River

San Manuel ■

Safford ■

Yuma ■

Willcox ■

Bowie ■

Ajo ■

Tucson ■

Benson ■

U.S.A.
Mexico

Sierra Vista ■

Douglas ■

Nogales ■

Bisbee ■

California

Arizona

Mexico

Baja California

The Sonoran Desert

The desert country

Land of wondrous contrasts: That's Arizona. In the north are the glorious Grand Canyon, scenic mountains and timberland. There are arid sections as well.

But it is the southern half of the state that is dominated by the remarkable Sonoran Desert. And it, too, has its contrasts.

Far more than parched land dotted with creosote and cacti, the Sonoran Desert has cooler grasslands that are shaded here and there by mesquite and palo verde trees, and disconnected mountain ranges that are never out of sight. Some have peaks that tower more than 9,000 feet and are snow-capped in winter.

The beauties of the Sonoran Desert are known and appreciated by many Americans, through the camera's eye if not in person. Visitors pour in each year by the thousands, and Western movies and television scenes are often shot against a backdrop of the desert's majestic saguaro cacti, rugged mountains and vivid sunsets.

The desert's size is impressive. As the map shows, it stretches far beyond Arizona. In dry years, its ragged, wedgelike shape grows larger than its normal 120,000 square miles. About a third of it is in the United States. Its name harks back to the time both sides of the border were part of the Mexican state of Sonora.

The desert reaches north of Phoenix, skirting the southern edge of the Mogollon Rim or Highlands. It curves southeastward around the Tucson area, bypassing Nogales on the west before it continues into Mexico. The Mexican two-thirds sprawls south on each side of the Gulf of California, deep into the states of Sonora and Baja California.

On the west, the Sonoran Desert extends beyond Yuma to the Indio area of California.

The Sonoran Desert portion of Arizona is laced with sunshine, and its growing seasons are long. Much of the Arizona portion is midaltitude desert (2,000 to 3,000 feet elevation). This includes the Tucson area in the southeast, and the area north and west of Phoenix.

In between is a strip of low-altitude desert (100 to 2,000 feet elevation) that stretches from Yuma to Phoenix.

The midaltitude enjoys 220 to 242 days without frost; the low-altitude, more than 300 frost-free days.

Rainfall is minimal. It comes in late summer and again in winter, reaching about 11 inches in the midaltitude desert. In the low-altitude strip, it measures less than 5 inches at Yuma, closer to 8 inches at Phoenix — in good years.

There are few rivers and streams in the desert. Some are dry except during the rainy seasons or when swollen with melted snow from the higher ranges. As the Colorado River flows toward Yuma, it furnishes vitally needed irrigation water. Among the handful of other desert rivers are the Salt, Gila, Verde and the Santa Cruz (which carries water above ground in the Tucson area for only part of the year).

Still, the desert is far from a Sahara. Longtime residents and newcomers alike are enthralled with the handsome yucca, ocotillo and other desert growth. Much of it has furnished food for native Americans for centuries.

In today's urban areas, people marvel at the variety of foods they can grow in their own yards with careful husbanding of precious water, a large portion of it from wells.

Bordering the Sonoran Desert on the east are high-altitude desert grasslands and mountains that are not actually a part of it, but are sometimes referred to as the "high" Sonoran Desert.

Some cacti and related plants grow in this high-altitude desert (elevation 3,300 to 5,000 feet), but not the Sonoran Desert's distinctive hallmark, the saguaro.

Introduction

When our decision was made to publish a "desert cookbook," it seemed only right that other Arizonans should participate.

And so The Star invited those interested to send us their favorite ways to use the fruits and nuts that grow so well in the Sonoran Desert and adjoining areas.

The response was far greater than we had anticipated, and the book was expanded to accommodate as many of the suggestions as possible.

The people who responded are as varied as their recipes. There is the Douglas resident who supplied us with the particulars for making elderberry and manzanita jelly. Her skills have pleased friends and neighbors for decades.

There is the retired University of Arizona professor who gave us the directions for her family's favorite fruit dessert.

There is the winner of a national recipe contest who, on the eve of her departure for the Pacific cruise she had won, sat down and wrote out for us the recipe for her lemon pies. She makes them often with lemons from trees at her Phoenix home.

And the teen-age Tucson boy whose cooking interests help him stay active despite a debilitating illness; and the longtime Wickenburg resident whose saguaro chiffon pies are justly famous; and the Western book writer who lives at Clifton, near the New Mexico border, whose method for drying apples with spices makes flavorful eating all winter.

A Tucson public relations executive, an enthusiastic solar-cooking advocate, shared her recipe for whole-grain-and-dried-apricot bread that can be baked conventionally as well as with sun power. Another working woman in Nogales gave us the directions for quince pastries, a popular Mexican dessert.

Women on ranches, rural housewives, urban women — and a sprinkling of men who also enjoy cooking — responded to the call for favorite recipes.

Some of the ideas are very modern. Others are old-style family favorites that come into use only at holiday times or family gatherings. All enrich this recipe collection. We wish there had been room for every one of them. The names of the people whose recipes do appear in this cookbook are on Pages 177 and 178.

For readers who are more armchair cooks than producers, the book touches on food history and lore. The book also calls attention to the astounding variety of edibles that thrive in our favorable climate, not only the fruits and nuts we cultivate in our own yards, but the native plants as well. These provided nourishment to the Native Americans who lived in this area for centuries.

The Papagos, Pimas, Maricopas and other American Indians of Arizona still use many of these native foods. For most of us, however, the wild fruits, berries and nuts are novelties.

As for the cultivated varieties, virtually every home in the Sonoran Desert and adjoining areas enjoys the beauty of one or more of the fruit-bearing trees that flourish in our climate. In the low and middle elevations, citrus, dates and olives are dominant. In the middle and high desert, apples and pears are successful. In all three areas, many people also raise grapes, melons and dozens of other fruits. Pecans and pistachios are among the nuts.

From the vantage point of The Arizona Daily Star's food desk, we see an increasing desire to add zest to eating by preparing some of the native fruits and nuts. People ask what they can do besides make jelly with the prickly pears that ripen in August and September, and cooking schools that show how to grind and use mesquite-bean flour draw avid students.

And people want to make better use of the fruits and nuts they grow in their own yards. Numerous calls and letters come to us throughout the year, asking for recipes for all these foods.

The book does not include recipes for tomatoes, squash, eggplant and peppers, even though they grow well here and, strictly speaking, are fruits. Tomatoes were declared a vegetable by the U.S. Supreme Court in 1893, and, like the other three, are found in cookbooks in the vegetable section. We follow the same procedure.

For those who do not raise or gather the foods covered in the book, they are often available from friends and neighbors or at the supermarkets.

Besides recipes, food lore and history, "Fruits of the Desert" provides some information on nutrition and on which varieties of fruits and nuts do best. In addition, there are sections on microwave cooking and on drying, a food-preservation method ideal in an arid land.

Acknowledgements

Friends are as important in professional life as in private. This is never truer than when doing a cookbook, I found. Everyone who was asked came to my aid with advice, information, encouragement and just plain hand-holding, when the need arose. My deepest appreciation to:

June C. Gibbs, state nutritionist, University of Arizona Cooperation Extension Service; George Brookbank, UA extension urban horticulturist; and county extension home economists for providing basic information as well as personal recipe favorites.

The UA College of Agriculture, the UA Arid Land Studies, Meals for Millions and the Arizona-Sonora Desert Museum for frequent assistance.

Elizabeth Shaw, assistant director, UA Press, and Phyllis W. Heald, free-lance writer and book consultant, for advice and hand-holding.

Ann Tinsley, UA School of Home Economics, who garnered for the cookbook the results of experimental food classes, and Hanna Lundberg and Gloria Thomasson, whose cooking expertise is well known in our area.

Home economists Mary Hardy and Kathryn W. Kazaros, for testing recipes and helping with decisions on which ones to use, and Carolyn Niethammer, book-writer and food historian, for assuring me that eventually it does indeed all come together.

And finally, my co-workers who so willingly assisted, into the late hours when necessary, in getting this book ready for printing.

— Sandal English

Contents

Contents

x

Berries

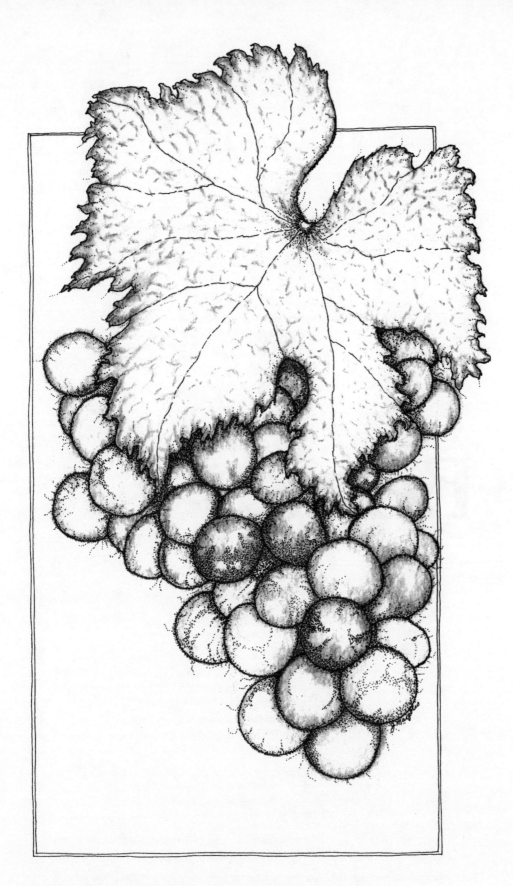

Grapes are among the berries that thrive in an arid climate.

Berries

Berries provide their own special flavor to homemade preserves or wine, or give zest to salads and desserts.

For Native Americans, wild berries have long added pleasure to eating. They used them fresh from the bush, stewed them or dried them for a sweet in winter. Sometimes hikers can still find these wild berries in arroyos, and foothill and mountain areas.

Both wild and cultivated berries offer fair amounts of vitamin C, especially when eaten raw. Some of the favorites in the Sonoran Desert and adjoining counties are included in this section.

Currants, gooseberries

Botanical name: *Ribes species*

Currants are somewhat acid berries that grow on shrubs seldom more than 6 feet tall. Though there are black varieties, the experts recommend a large red berry called Perfection for growing in Arizona at 3,500 to 7,000 feet.

These berries are not the same as the black currant grown in northern Europe and Asia and used to make the excellent liqueur called cassis. (To make matters somewhat confusing, a small dried seedless grape used in cooking also bears the name "currant." The grape's name is derived from the fact that the raisins came from Corinth; hence their name, *raisins de Corauntz*.) Wild currants are found from 5,000 to 9,000 feet, in moist areas near other trees and bushes. Red in color, they are not very juicy.

A close relative of currants, gooseberry shrubs prefer cooler, higher elevations. The berries may be yellow, green or whitish, and usually are harvested before fully ripe, for jam and jelly making.

They are grown commercially in Colorado and some of the other Western states, but because they — as well as currants — can bear a fungus that causes white pine blister rust, their planting is forbidden in some parts of the country.

For cultivating at 3,500 to 7,000 feet, the best gooseberry varieties are Pixwell, Carrie and Champion.

Currant jelly

*From her backyard bushes, **Ora Sigrist** of Douglas has plenty of berries in midsummer for making jelly.*

3 quarts currants
3 quarts water
6 cups sugar
1 bottle liquid or 2 packets pectin

Remove stems and wash berries well. Place in a 6-quart kettle, add water and cook over medium heat for 30 minutes. Drain in colander that has been placed over a large bowl or kettle to collect the juice. Strain juice through a jelly bag or several thicknesses of cheesecloth.

Measure 4 cups juice into 6-quart kettle. Bring to rolling boil and stir in 3 cups sugar. Bring back to boil, stir in remaining sugar and the pectin. Bring back to a full rolling boil (a boil that cannot be stirred down). Continue cooking until jelly sheets from a large metal spoon. Pour into hot sterile glasses. Add paraffin or seal at once. Makes about 6 8-ounce glasses. Repeat with remaining juice, for another batch of about 6 glasses.

Gooseberry fool

This old-fashioned dessert is simple enough to delight modern cooks.

2 pints gooseberries
¾ cup sugar
1 cup whipped cream

Cook gooseberries in a saucepan with the sugar about 15 minutes, or until soft. Put through a sieve. If there is a lot of juice, set some aside. Cool fruit. Mix the whipped cream and refrigerate until ready to serve. Serves 6.

Bread-and-butter pudding

Dried currants and brandy are layered with bread slices for an old-fashioned pudding.

¾ cup dried currants
2 tablespoons brandy
20 slices untrimmed French bread, each about ⅓-inch thick
3 egg yolks
3 whole eggs
1 cup sugar
2 cups milk
1 cup heavy cream
¼ cup melted butter
Confectioners' sugar

Mix currants with brandy and set aside for 30 minutes. Slice the bread. (Cut slices in half if larger than 3 inches in diameter.) Combine the egg yolks, whole eggs and sugar in a mixing bowl and beat until blended. Stir in the milk and cream. Drain the brandy from the fruit and add the brandy to the egg mixture.

Spread fruit over the bottom of an oval baking dish. Brush 1 side of each bread slice with butter. Arrange the bread slices neatly, overlapping.

Strain the custard over the bread slices.

Set the baking dish in a larger baking dish with 1 inch boiling water in it. Bake at 350 degrees for 40 minutes to 1 hour, until custard is set.

Cool on rack. Sprinkle with confectioners' sugar. Serves 10 to 12.

Elderberries

Botanical name: *Sambucus mexicana et al*

Elderberries grow on small semi-evergreen trees or bushes found along streams and ditches from 1,000 to 4,000 feet elevation. The edible, dark blue berries ripen in summer. They are sweet and juicy, and can be used to make jelly. Elderberry wine is considered a great delicacy.

In urban settings where they are watered, elderberry trees can grow to 20 feet, with a spread almost as wide. They flower from winter to late spring. The fruit ripens from spring (in the lower levels) to midsummer (in higher levels).

It was long believed that bad luck would follow if elder trees were uprooted. Both elderberries and flowers were considered medicinal (and sometimes are still used in this fashion). The juice of the berries "cleansed" the blood and relieved chills during winter; elder-flower tea, once a treatment for colds and other ailments, is a flavorsome drink either hot or cold.

Elderberry jelly

Standard directions call for covering berries with water, simmering until tender, mashing and straining through a jelly bag.

Put ½ package powdered pectin in a large kettle with 2 cups juice; stir and bring to a boil. Stir in 2½ cups sugar and 2 tablespoons lemon juice. Bring to a boil and boil hard for 2 minutes. Skim and pour into jelly glasses and seal.

Grapes

Botanical name: *Vitis labrusca (American grape); V. vinifera (European grape); V. arizonica (wild grapes)*

What comes red, purple or pale green and is loved by everybody? Grapes, of course. Those not eaten fresh can be dried into raisins or made into juice, jelly or wine.

These time-honored ways of using grapes come to us from ancient days. It's written that the Garden of Eden had grapes. For untold ages, those who lived in the warm climates near the Mediterranean enjoyed them.

Love for grapes is well-established in the Sonoran Desert. They were introduced to the area when missions were established in the 1700s, and our dry, hot weather proved ideal. Today, commercial viticulture (grape-growing) is developing in Arizona, as a result of research that catalogs good producers, some even on substandard land.

For home-growing, where temperatures do not drop below 10 degrees, you'll find European types, such as Cabernet Sauvignon and Cardinal popular among the reds; Perlette, Thompson Seedless and White Riesling among the whites.

Elevations about 5,500 feet can grow the Concord, as well as the reddish Caco, purple Fredonia and white Niagara. These four make excellent jellies or dessert wines.

Grapes require careful pruning to ensure full, heavy bunches. When trellises of the vines are installed close to houses, the vines can provide seven to eight degrees of cooling in the hot desert summers.

Wild grapes grow along streams and canyons from 2,000 to 7,000 feet.

Raisin fruit soup

Tucsonan **Jane Nyhuis** *serves this soup hot in winter, chilled in summer, topped with plain yogurt.*

1 cup raisins
1 cup dried apricots
1 cup canned peaches, diced, with juice
1 can pineapple chunks and juice
1 orange, peeled and diced
½ cup almond liqueur
  Plain yogurt

Combine fruits and juices and simmer for about 20 minutes. Stir in ½ cup almond liqueur and 1 teaspoon cinnamon. Serve hot or cold, each serving topped with 1 tablespoon yogurt. Serves 4 to 6.

Fruit-and-nut chicken salad

This salad is one of Tucsonan **Betty Crane's** *favorite bridge-luncheon dishes.*

1 cup mayonnaise or salad dressing
  (Miracle Whip or other)
2 tablespoons soy sauce
⅔ cup sour cream
1 teaspoon paprika
2 cups seedless grapes
4 cups cooked chicken cut in bite-size
  pieces
2 cups finely chopped celery
1 cup orange segments, quartered
  Salt and white pepper to taste
  Lettuce, fruits
1 cup whole toasted almonds

Combine mayonnaise with soy sauce, sour cream and paprika, and toss lightly with grapes, chicken, celery, orange segments, salt, pepper. Serve on bed of lettuce surrounded by fruits and garnish with toasted almonds. Serves 6 to 8.

Grape-nectarine yogurt salad

This gelatin fruit salad features grape halves and yogurt, with nuts and nectarines.

2 envelopes unflavored gelatin
1½ cups fresh orange juice, divided
1 tablespoon sugar
2 8-ounce containers plain yogurt
3 tablespoons honey
½ teaspoon vanilla
¾ teaspoon grated lemon or orange rind
1½ cups diced, peeled nectarines or pears
1½ cups halved grapes, seeded
¼ cup chopped walnuts or almonds

In medium saucepan, sprinkle unflavored gelatin over 1 cup orange juice. Place over low heat. Stir until gelatin dissolves, about 3 minutes. Remove from heat. Stir in sugar, remaining ½ cup orange juice, yogurt, honey, vanilla and lemon rind. Stir until mixture is smooth. Chill, stirring occasionally, until mixture is consistency of unbeaten egg whites. Fold in nectarines, grapes and nuts. Turn into a 2-quart bowl or 8 dessert dishes and chill until set. Serves 8.

Tossed grape salad

Seedless grapes combine nicely with spinach for a salad to be served with bacon dressing.

6 cups lightly packed torn spinach leaves
6 ounces Swiss cheese
6 ounces Cheddar cheese
2 cups seedless grapes
4 slices bacon
¼ cup cider vinegar
½ cup oil
2 teaspoons Dijon-style mustard
1 teaspoon seasoned salt
¼ teaspoon pepper
2 tablespoons sliced green onion

Arrange spinach, cheeses and grapes on 4 individual serving plates. Top with bacon dressing, which is prepared this way: Cook bacon until crisp, drain and crumble. In small bowl, combine vinegar, vegetable oil, mustard, seasoned salt and pepper. Stir in bacon and onion. Cover and chill several hours or overnight. Just before serving, mix thoroughly.

Grape-pecan chicken salad

Mary Love Burks *of Tucson says this chicken salad recipe from her Texas mother-in-law makes "heavenly eating."*

1 broiler-fryer chicken
    Bayleaf, salt, pepper
1 bunch green onions, chopped
1 cup chopped celery
1 cup seedless grapes, halved
1 cup chopped pecans
½ cup mayonnaise
2 teaspoons sugar (about)
1 tablespoon lemon juice
    Salad greens

Simmer fryer with bay leaf, salt and pepper. Remove chicken from bone and cut in bite-size pieces. Combine chicken with onion, celery, grapes and pecans. Combine mayonnaise, sugar and lemon juice. (Use less than 2 teaspoons sugar if grapes are quite sweet). Chill. Serve on salad greens to 8, or use as a sandwich spread.

Cinnamon rolls

Ella Filer *of Tucson considers baking a creative art; in fact, she calls it "interior decorating," and these rolls seem to prove her point.*

4 cups flour
3 tablespoons softened margarine
2 beaten eggs
1 cup raisins
1 teaspoon salt
1 package dry yeast softened in 1 cup
    warm water
¼ cup sugar
1 tablespoon oil
½ cup brown sugar mixed with 1 teaspoon
    cinnamon

In large bowl mix the flour, margarine, eggs, raisins, salt, softened yeast and ¼ cup sugar. Knead until a soft dough is formed. Let rise until double in bulk in greased, covered bowl. Roll out in large rectangle and brush with oil and sprinkle with sugar-cinnamon mixture.

Roll up as for jelly roll and cut into 1-inch slices. Place slices in large greased pan or in muffin tins. Let rise again. Bake for 15 to 20 minutes in 350-degree oven.

Note: If desired, top while rolls are still warm with a thin mixture of powdered sugar and orange juice.

Green grape pie

In this grape pie version, sent by **Nancy L. Wilson** *of Globe, sour cream and lemon juice add their special flavors. The recipe is from* **Phyllis Wells** *of Casa Grande.*

    1 baked graham-cracker crust
    1 quart green seedless grapes
    ¾ cup sugar
    3 tablespoons cornstarch
    ¼ cup cold water
    1 tablespoon lemon juice
    1 cup sour cream
    1 tablespoon sugar
    1 teaspoon vanilla

   Combine grapes and ¾ cup sugar in pan. Dissolve cornstarch in cold water and gently stir into grape mixture. Bring to a boil, stirring gently, then reduce heat and simmer about 5 minutes. Remove from heat, stir in lemon juice. Cool. Turn grape filling into pie shell. Blend sour cream with 1 tablespoon sugar and vanilla and spread evenly over top. Sprinkle top with reserved crumbs. Serves 6.

Fresh grape pie

Seedless grapes are used by **Hazel M. Battiste** *of Tucson to make this spiced grape pie.*

    Pastry for lattice-top pie
    2 tablespoons instant tapioca
    1 teaspoon cinnamon
    ¼ teaspoon nutmeg
    1½ to 2 cups sugar
    4½ cups seedless grapes, cut in halves
    2 tablespoons butter
    2 tablespoons water

   Mix the tapioca, cinnamon, nutmeg and sugar. Toss grapes in sugar mixture and fill pie shell. (It will be full, but grapes will flatten as they cook.) Dot with butter and add water. Cover with pastry strips and flute

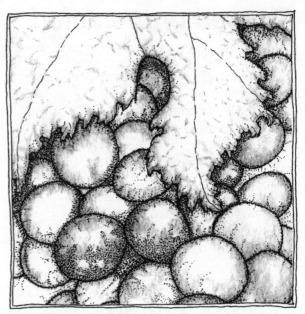
Grapes

pie edge high to hold juices. Cover edge of pie with strips of foil to prevent burning. Bake at 425 degrees for 15 minutes then at 325 for 35 to 40 minutes. Cool before serving.

Raisin vinegar pie

Ruth Brinkerhoff *of Safford enjoys making this pie for special occasions, especially during the holidays.*

    2 cups raisins
    2 cups cold water
      Pinch salt
    2 eggs
    2 cups sugar
    4 tablespoons flour
    4 tablespoons melted butter
    4 tablespoons cider vinegar
      Pastry for 2-crust pie

   Boil raisins in water until tender. Drain. Add cold water and other ingredients. Cook and stir until thickened. Cool. Transfer to filled pie plate and put on top crust. Slash and bake at 450 degrees for 10 minutes, then at 350 for 30 minutes or until brown.

Kentucky jam cake

*Raisins join jam in this old family favorite from **Kathryn Rapp** of Bisbee. It used to be made often by her mother and makes a good substitute for fruitcake at holiday-time.*

    4 cups flour
    1 teaspoon soda
    ½ teaspoon salt
    2 teaspoons of a combination of cinnamon,
        allspice, nutmeg and cloves
    1 cup butter or margarine
    2 cups sugar
    4 eggs
    1 cup berry jam
    1 cup peach or apricot jam
    1 cup buttermilk
    ½ cup raisins or dates
    1 cup chopped pecans

Sift together flour, soda, salt and spices. Cream butter, gradually adding sugar. Cream until light and fluffy. Beat in eggs, one at a time. Add jam and beat. Add dry ingredients and buttermilk. Fold in chopped raisins and nuts, mixing only until well distributed. Pour into 3 wax paper-lined 9-inch cake pans. Bake at 350 degrees for 30 to 45 minutes. Frost with caramel frosting.

Snow white fruitcake

Marjorie T. Trough *of Tucson shares this version of a 100-year-old recipe from her grandmother.*

    1 cup butter
    1½ cups sugar
    4 cups flour
    1 teaspoon salt
    1 teaspoon baking powder
    ½ teaspoon nutmeg

    1½ cups white raisins
    2 8-ounce jars candied fruitcake mixture
    10 egg whites, beaten stiff
    1 cup cream sherry

Cream butter and sugar. Mix sifted flour, salt, baking powder and nutmeg in large bowl. Mix in raisins and candied fruit. Combine with butter-sugar mixture and the sherry. Fold in egg whites. Bake in 2 paper-lined loaf pans for 2½ hours at 250 degrees, then increase temperature to 300 degrees for 15 more minutes. Cool on rack and remove from pans. Wrap each loaf in a linen napkin and saturate with cherry wine for 2 weeks before Christmas. Store each loaf in a tin box, adding more wine if desired.

Pinto bean and raisin cake

*This recipe, sent in by **Judy Diers** of Tucson, is an old-fashioned pinto bean cake with raisins and nuts.*

    1 cup sugar
    ½ cup butter or margarine
    1 egg, beaten
    2 cups cooked pinto beans (cooked with
        about 1 teaspoon soda) and mashed
    1 cup all-purpose flour
    ½ teaspoon salt
    1 teaspoon cinnamon
    ½ teaspoon ground cloves
    ½ teaspoon allspice
    1 cup raisins
    2 cups diced raw apples
    ½ cup chopped nuts
    2 teaspoons vanilla

Cream sugar and butter; add egg and beans. Sift flour, salt, cinnamon, cloves and allspice and add to first mixture. Fold in fruits, nuts and vanilla. Pour into well-greased loaf or tube cake pan and bake at 375 degrees for 45 minutes. Top with light glaze, if desired. Note: if using beans cooked without soda, sift about ¾ teaspoon soda with dry ingredients.

Grape fruitcake

Fresh grapes join apples and nuts to make this interesting fruitcake.

1 cup sugar
¼ cup butter
1 teaspoon cinnamon
¼ teaspoon nutmeg
1 teaspoon vanilla
2 cups coarsely grated apples
1½ cups grapes, halved and seeded
½ cup chopped nuts
1 cup flour
1 teaspoon baking soda

Cream sugar, butter, cinnamon and nutmeg in mixing bowl. Stir in vanilla, apples, grapes and nuts. Combine flour and baking soda; sift into creamed mixture and blend well. Spread batter in lightly greased 8-inch square cake pan. Bake at 350 degrees for 40 to 45 minutes or until cake tests done. Glaze with lemon icing. Serves 8.

Lemon icing: Combine 1 cup sifted powdered sugar with 1 tablespoon grated lemon rind and ¼ cup lemon juice. Beat until smooth.

Chewy raisin cookies

Back in Michigan, Grandma Warner raised her children on these cookies, and so did **Ruth Warner,** *who now makes this family favorite with raisins dried from the Thompson seedless grapes she grows in her Tucson yard.*

2 cups sugar
½ cup margarine or butter
½ cup vegetable shortening or lard
3 eggs
2 to 3 tablespoons milk
1 cup ground raisins
1 teaspoon soda
1 teaspoon cinnamon

¼ teaspoon nutmeg
 Pinch of salt
4 cups flour (about)
 Chopped nuts (optional)

Cream sugar, margarine and shortening and beat in eggs. Add milk and work in raisins. Mix dry ingredients and add to first mixture. Add nuts. Roll out dough on lightly floured surface to ¼ inch thick. Cut with cutter and bake on lightly greased cookie sheets. For a 2½-inch cutter, use 375 degrees for about 12 minutes. Makes 4 dozen chewy cookies.

Oatmeal-raisin cookies

No cookie was ever better than a moist, raisin oatmeal cookie like this one from **Anna Marie Mollison** *of Tucson.*

1 cup seedless raisins
1 cup shortening
1 cup sugar
3 beaten eggs
2 cups flour
½ teaspoon each salt, soda, allspice and cloves
1 teaspoon cinnamon
2 cups quick oatmeal (uncooked)
6 tablespoons raisin liquid
½ cup chopped dates
1 cup broken pecans

Cook raisins in boiling water to cover for 5 minutes. Drain, reserving 6 tablespoons juice. Thoroughly cream shortening with sugar; add eggs and beat until smooth. Sift flour with salt, soda and spices. Mix with oatmeal and add to creamed mixture alternately with raisin liquid. Add dates and nuts. Drop onto greased cookie sheets and bake at 400 degrees for 10 to 12 minutes. Makes about 4 dozen.

Fruit dessert-salad

A popular salad with **Margie Valentine** *of Safford is this old-style fruit mixture, a recipe from her mother who lives in Texas.*

2 cups seedless grapes
2 unpeeled apples, cubed
2 oranges
2 bananas
2 peaches, peeled
2 cups pecans
1 cup grated coconut
½ cup evaporated milk
1 tablespoon sugar

Mix fruits, nuts and coconut. Combine milk and sugar and add to fruits. Refrigerate until serving. Serves 12 or more.

Grape-pear chutney

Raisins combine with pears to make this exotic chutney, which can be stored in the cupboard in sterilized jars or unsealed in the refrigerator.

3 pounds fresh pears
1 pound brown sugar
1 pint cider vinegar
1 medium onion, chopped
1 cup raisins
¼ cup diced preserved ginger
1 clove garlic, minced
½ teaspoon cayenne or red pepper
2 teaspoons salt
½ teaspoon ground cinnamon
½ teaspoon ground cloves
2 teaspoons mustard seed

Core and dice unpeeled pears. Combine brown sugar and vinegar and bring to a boil. Add pears and remaining ingredients. Cook slowly, stirring from time to time, until thick, about 1 hour. Pour into hot, sterilized jars and seal. Or store unsealed in refrigerator, for 3 to 4 weeks. Makes about 5 ½-pints.

Dried fruit treat

For healthful, high energy snacking, nothing can beat dried mixed fruits and nuts, **Trudy Lucas** *of San Manuel believes.*

2 cups raisins
1 cup grated coconut
1 or 2 cups dry-roasted peanuts
1 cup toasted sunflower seeds
Dried banana slices

Combine ingredients and salt lightly before bagging. To dry bananas, slice and spread on cookie sheet and dry in 150-degree oven for 9 hours, turning once.

Mock cherries

Betty J. Faris *of Globe was given this recipe for using home-grown Thompson seedless grapes when she moved to Arizona in the late '60s. Her husband, too, enjoys eating them.*

Thompson seedless grapes for 6 pints
1 package cherry-flavored unsweetened drink mix
Liquid sweetener to equal 1 cup sugar
1 quart cold water
1 teaspoon or less red food coloring

Wash grapes and pack into clean pint jars. Combine drink mix, sweetener, water and coloring. Stir until dissolved. Fill jars of grapes with the mixture, leaving ½-inch head space. Screw on lids and process for 15 minutes in boiling water bath. Remove and cool. Store at least a month before using as canned cherries for cherry pie or crisp. Almond flavoring or nutmeg may be added.

Note: It is better to use sweetener instead of sugar, because the sugar reacts with the tartaric acid in the grapes and forms unwanted crystals.

Mulberries

Botanical name: *Morus nigra*

Resembling miniature blackberries, mulberries are sweet to the taste, but it's a scramble to get to them before the birds do!

Both the fruit-bearing (female) and fruitless (male) are imports to the Sonoran Desert and are grown primarily for their excellent shade. Once established, the trees do well with little water. But there are drawbacks to raising mulberries. Though they have no pollen, the fruiting female trees tend to leave a mess when their fruit drops to the ground. On the other hand, the fruitless male trees pour out pollen in a tremendous flow in spring, becoming a major source of torment to allergy sufferers.

Using mulberries

For jelly, simmer berries in water until tender. Drain through jelly bag. Use 1 part juice to 1 part sugar and boil to jelly stage. For jam, simmer berries in own juice until tender. Add equal amount of sugar and cook until thick.

Pyracantha berries

Botanical name: *Pyracantha coccinea, P. fortuneana, P. koidzumii*

The pyracantha is also known as "firethorn," a fitting name. The bush is a thorny evergreen shrub whose fragrant white spring flowers are followed by bright clusters of red or orange berries. Some varieties produce berries in late summer, others in the fall. They remain on the bushes for many months, adding bright touches to yards until Christmas. The birds love them.

Pyracantha bushes can be espaliered or trimmed in various shapes for landscape use. They require little water, but their pollen creates a serious breathing problem for those with allergies.

The berries are often turned into jelly after adding lime or lemon juice to enhance flavor. Pyracantha wine is an attractive pale gold.

Pyracantha jelly

Iola Martin *of San Manuel shares this method for making jelly from berries of the pyracantha bush.*

Wash 1 pint ripe berries (for deeper color, use 1 quart) and boil in 3 pints water for 20 minutes; strain berries through cheesecloth. To 3½ to 4 cups juice, add ¼ cup lemon juice, ¾ cup grapefruit juice and 1 package powdered pectin. Bring to a rolling boil, stirring. Add 5½ cups sugar and boil and stir for 2 minutes. Remove from heat, skim and pour in sterilized jars and seal. Caution: Do not double recipe.

Raspberries

Botanical name: *Rubus species.*

This delightful berry is available fresh for only a few weeks each summer at the supermarket; most of the commercial crop is sold frozen.

Raspberries do best in home gardens at 3,500 to 7,000 feet, and those who grow them find their distinctive flavor well worth the pruning and other care required. Among the recommended varieties are Latham, Red Chief and Newberg, all reds; and Cumberland, purple.

Perhaps because of their high vitamin C content, raspberries were long used medicinally. Raspberry-leaf tea was used as a preventive to miscarriage and raspberry vinegar was said to relieve sore throat.

Relatives of the raspberry — **blackberries, boysenberries and loganberries** — also can be grown successfully in home gardens.

Blackberry dumplings

Elizabeth Ferguson of San Manuel, who grows several kinds of berries in her yard, makes this dessert with her blackberries.

1 cup sifted flour
1¼ teaspoons baking powder
1 teaspoon salt
2 tablespoons sugar
3 tablespoons shortening
½ cup milk
1 quart ripe blackberries
1 cup sugar
¼ cup water

Sift flour, baking powder, salt and 2 tablespoons sugar together. Cut in shortening with pastry blender to make coarse crumbs. Add milk; mix to make soft dough. Combine berries, 1 cup sugar and water. Bring to a boil. Drop dough by tablespoonfuls onto the boiling fruit. Simmer, uncovered, for 10 minutes. Cover; simmer 10 minutes more. Serve with cream or ice cream. Serves 6.

Berry pudding

Raspberries, strawberries, red currants or other berries may be used in this pudding.

8 slices stale, crustless white bread,
   ½ inch thick
1½ pounds raspberries and other berries
   ½ cup sugar or more, depending on berries

Line the bottom of a 4-cup soufflé dish with enough bread to cover the base completely. Line the sides of the dish with more bread slices; if necessary, cut them to shape so that the bread will fit closely together. Hull and carefully wash the fruit. Put the fruit in a wide, heavy-bottomed pan and sprinkle sugar over it. Bring to a boil over very low heat; cook for 2 to 3 minutes only, until the sugar melts and the juices begin to run.

Remove the pan from the heat and set aside 1 to 2 tablespoons of the fruit juices. Spoon the fruit and the remaining juice into the bread-lined soufflé dish, and cover the surface completely with the rest of the bread.

Put a plate that fits the inside of the dish on top of the pudding, and weight it down with a heavy object. Chill the pudding in the refrigerator for 8 hours. Before serving, remove the weight and plate. Cover the dish with the serving plate; hold the plate firmly against the dish and turn upside down to unmold the pudding. Use the reserved fruit juice to pour over any parts of the bread that have not been soaked through and colored by the fruit juices. Serve the pudding with a bowl of cream or whipped cream. Serves 6.

Raspberry sauce

The pleasure of eating fresh raspberries is enhanced in this recipe with lemon juice and orange liqueur.

   1 pint fresh raspberries
    Juice of half a lemon
¼ to ½ cup sugar
   2 tablespoons orange liqueur
    Vanilla ice cream or fresh strawberries
      or raspberries

Rinse and drain raspberries. Put in food processor or blender and add the lemon juice and sugar. Blend thoroughly.

Add liqueur and blend. Serve with vanilla ice cream or over whole fresh fruit such as fresh strawberries or fresh raspberries. Makes about 1½ cups.

Blackberry-apple pie

Boysenberries or loganberries can be substituted in this recipe.

    Pastry for 1-crust pie
   2 pints blackberries
   3 cooking apples
¾ cup sugar or more, depending
      on tartness of fruits
   1 egg white

Wash blackberries; core and slice apples. Arrange fruit in layers in buttered pie dish, sprinkling each layer with a little sugar.

Roll out the pastry to ¼ inch thickness on a floured board. Cover the pie with the pastry, allowing about an inch to overlap the sides. Seal the sides down with the prongs of a fork. Decorate the crust with leftover pastry pieces cut in the shapes of leaves. Glaze with egg white beaten with a little water.

Bake at 425 degrees for 15 minutes, then turn heat down to 350 degrees and bake for another 30 minutes.

Strawberries

Botanical name: *Fragaria species*

Herbalists consider tea made from strawberry leaves valuable in relieving aches and pains, but for most of us, strawberry shortcake and other pleasurable dishes with strawberries are sufficient reason to grow them.

The name strawberry likely derives from the old practice of laying straw under the plants to prevent the ripe berries from becoming moldy.

A member of the rose family, strawberries require plenty of room and good, rich soil. They also grow quite successfully in space-saving wooden barrels or tubs. Arizona has quarantine regulations against importing strawberry plants from states where the strawberry root weevil is a pest. County cooperative extension agents can provide up-to-date information.

In the Salt River and Gila River valleys, June-bearers include Lassen and Tioga; in the Tucson area, Sequoia, Tioga, Frensa and Quinault are good choices. The varieties that grow well at 3,500 to 7,000 feet include Ozark Beauty, Ogalala and Sequoia.

Strawberries

Strawberry soup

For an unusual fruit appetizer, try this fruit soup. It can be served warm or cold.

1 quart ripe strawberries, hulled and rinsed
½ cup sugar (or more to taste)
  Pinch salt
1 cup sour cream
1 cup dry red wine
4 cups cold water

Place all ingredients except water in a blender or food processor. Process until smooth. Combine strawberry purée with water in saucepan; heat slowly until hot. Serve warm or cold. Serves 8.

Strawberry gelatin salad with nuts

If available, fresh strawberries from her son's garden are used when Betty Thomas of San Manuel makes this salad. Otherwise, she uses strawberries from her freezer.

1 No. 2 can crushed pineapple
1 family-size package strawberry gelatin
1 cup fresh (or frozen) strawberries, sliced
½ cup chopped nuts
1 pint sour cream

Drain pineapple, reserving juice. Add enough liquid (other fruit juice or water) to make 1½ cups. Heat in pan and dissolve gelatin in liquid. Then stir in remaining ingredients except sour cream. Pour half of mixture into rectangular dish and chill until partially set. Spread sour cream over first layer and top with remainder of mixture, chilling until firm. Serves 12.

Glorified rice dessert

Flavorful and nutritious too, this rice dessert from Susan Martin of Thatcher was made to be enjoyed by brown-rice enthusiasts.

1 cup brown rice
2½ cups water
½ cup strawberries
½ cup cherries
½ cup raisins
½ cup pineapple chunks, drained
½ cup chopped pecans
½ cup whipped cream topping

Steam brown rice in water, covered, for 45 minutes or until done. Cool. Fold in fruits. Chill. At serving time, top with whipped cream. Serves 6 or more.

Strawberry-orange fruit ice

Strawberries and citrus juices combine for this easily made fruit ice.

2 pints red, ripe, sweet strawberries
1½ cups sugar
¼ cup lemon juice
2 cups orange juice

Remove the hulls from the strawberries. Trim away any blemishes. Slice the strawberries and put them into the container of a food processor or blender. Blend to a fine purée. Add the sugar and lemon juice. Blend briefly. There should be about 4 cups. Put the mixture into a mixing bowl and add the orange juice. Pour into the container of an electric or hand-cranked ice cream freezer and freeze according to manufacturer's instructions. Serves 8.

Chocolate-covered strawberries

Better than you can imagine are these fresh strawberries coated in chocolate.

1 pint strawberries
4 ounces semisweet chocolate or 1 6-ounce package semisweet real chocolate bits
Wooden skewers

Wash berries and remove stems. Dry on paper toweling. Insert skewer in stem end of each berry. Break chocolate in pieces and melt in top of double boiler over simmering water, stirring constantly. Remove as soon as chocolate is melted. Swirl each berry into the chocolate, covering most of berry, then stand the skewered berries in a drinking glass to dry.

Dipped berries also may be chilled in refrigerator to speed up hardening of the chocolate. Will keep in refrigerator up to 24 hours. Makes 1 pint.

Strawberry sauce

*One cup ripe peaches, chopped, can be substituted for the strawberries in this recipe from **Ellen Young** of Thatcher.*

Purée in blender 1 cup ripe strawberries, 1 cup pineapple juice and 1 tablespoon cornstarch. Pour into pan and simmer until thickened, stirring constantly. Use over pancakes or waffles. Makes 1½ cups.

Strawberry-orange conserve

Honey is part of the sweetening in this delicious conserve.

1½ pints fully ripe strawberries
1 medium orange
¼ cup finely chopped pecans
1 cup honey
3 cups sugar
¾ cup water
1 1¾-ounce package powdered fruit pectin

Stem and thoroughly crush the strawberries, working with a layer at a time. Measure 1½ cups into a large bowl or pan. Section orange, discarding peel. Dice orange sections and measure ½ cup. Add orange sections and nuts to strawberries.

Thoroughly mix honey and sugar into fruit; let stand for 10 minutes. Mix water and fruit pectin in small saucepan. Bring to a boil and boil for 1 minute, stirring constantly. Stir into fruit. Continue stirring for about 3 minutes. Ladle quickly into scalded, drained containers (1 pint or less in size). Cover with tight lids. Let stand at room temperature for 24 hours to set, then store in freezer. If to be used within 2 or 3 weeks, store in refrigerator. Makes 6 cups.

Tomatillos

Botanical name: *Physalis ixocarpa, P. pruinosa and others*

Tomatillos resemble cherry tomatoes, except for their papery husks. (These are removed before using.) In Mexican cooking, green tomatillos go into *salsa* (sauce).

Also known as ground cherries, husk tomatoes or strawberry tomatoes, the seedy fall berries are yellow to purple when ripe, depending on species. Some people find the ripe, sweet berries rather insipid; others describe the flavor as resembling a combination of cherries and strawberries. A kindred fruit, popularly known as Chinese lanterns, is inedible.

Ripe ground cherries (seeds and skins, too) may be eaten out of hand, served with sugar and cream or made into pie, jam or jelly.

Tomatillas grow on leafy, vining plants that can be cultivated in home gardens. The berries sold in supermarkets can be as large as golf balls. The smaller wild ones grow on medium-dry to moist ground, at elevations up to 8,500 feet.

Tomate verde con queso

Connie Brice of Phoenix sends this skillet dish made with tomatillos and cheese and seasoned with cilantro (coriander).

2 tablespoons butter
1 medium onion, finely chopped
1 clove garlic
4 or 5 good-sized tomatillos, chopped
 and simmered in small amount of water
 for 15 minutes
1 cup peeled and chopped green chiles
6 sprigs fresh cilantro, chopped
 Salt and pepper to taste
½ pound cheese cubes
½ cup evaporated milk

Heat butter in skillet; add onions and garlic and sauté until onion is tender. Add tomatillos, chiles, cilantro, and salt and pepper to taste. Simmer gently for 10 minutes, uncovered. Add cheese and when it begins to melt, stir in milk and heat through.

Tomatillo pie

When tomatillo cherries are ripe, they can be made into a pie.

Pastry for 2-crust pie
4 cups fresh ripe ground cherries
1 cup sugar (about)
3 tablespoons flour
¼ teaspoon cinnamon
⅛ teaspoon nutmeg
1 tablespoon butter

Line pie plate with crust. Combine sugar, flour and spices and stir into cherries. Pour into pie crust and dot with butter. Cover with top crust, slash cover and bake at 450 degrees for 10 minutes. Reduce heat to 350 and bake until pie is golden brown (about 30 more minutes).

Tomatillo salsa

*Tucsonan **Lorraine Moreno** makes this salsa for her husband, who learned to love it when he lived in his native Mexico.*

1 pound medium-size green tomatillos
10 to 12 small yellow hot chiles
2 or 3 green onions
1 clove garlic
    Salt to taste

Stew tomatillos in a little water until soft. Put in blender with chiles, onions (minus the green tops), garlic and salt, and blend well. Return to stove and bring to a boil. Store in refrigerator.

Other berries

Desert hackberries

Botanical name: *Celtis pallida*

A dense, spiny small tree or evergreen shrub, the desert hackberry has tiny flowers that are followed by small orange berries in the fall.

The tree can grow to about 10 feet high, and furnishes excellent cover for quail on desert plains. The tree also grows in dry washes and canyons at 2,000 to 4,000 feet. For the urban yard, desert hackberry makes an informal plant that helps control soil erosion. It will grow from seeds.

Though mainly of interest to birds, the one-seeded berries are suitable for making jam (see mulberry directions) and fruit sauce, or they can be dried for a sweet snack.

Juniper berries

Botanical name: *Juniperus species*

The plant is an evergreen that produces small purplish berries. It comes in dozens of species, some of which are ground covers, others of which grow into trees. In the home garden, junipers are among the most serviceable of landscape plants. The plants grow wild at 3,000 to 8,000 feet and like the company of the piñon pines.

Some varieties of the berries are dried for use in cooking. The flavor is somewhat sweet and has an appealing perfume that smacks of pine. Perhaps the most famous use of juniper berries is to flavor gin. Those grown as ornamentals in urban yards need sampling for palatability.

Hikers in the Graham, Chiricahua and Huachuca mountains sometimes can find the wild berries and bring home a few handfuls to flavor their cooking. When fully mature, they should be purple or reddish and taste slightly sweet. They may need to be dried before using.

Manzanita berries

Botanical name: *Arctostaphylos species*

This small, apple-shaped red berry carries the Spanish name that means "little apple." A shrubby evergreen with smooth, reddish bark on its crooked branches, it grows wild above 3,500 feet and can be grown in urban yards.

The berry is also called "bearberry." Though Native Americans have long eaten the berries raw, they are used by others mainly for making jelly.

Rose hips

Botanical name: *Rosa rugosa and R. multiflora, cultivated; R. arizonica, wild*

Rose hips are the seed pods that appear in the fall after the rose has bloomed and the petals have fallen. (If the flowers are cut, the hips do not develop.)

Rose hips are an outstanding source of vitamin C. The wild or dog rose that grows in higher elevations is especially rich in the vitamin. In urban settings, root stocks that bloom are more likely to produce rose hips than grafted roses. The hips come in sizes ranging from pea-size to marble-size and are generally red when ripe. Rose hips can be made into tea, jam and other foods, and doing so was popular with past generations.

Rose petals are an old-fashioned ingredient in cooking that is enjoying a revival. A distillation of rose petals becomes rose water and this too is used in cooking.

Using wild juniper

Martha "Muffin" Burgess of Tucson offers two suggestions for using wild juniper berries.

Stuffing: Add 10 to 15 ripe berries to your favorite cornbread stuffing for duck or turkey, along with diced green apple. Baste with apple cider. Gives a natural, woodsy flavor to the bird.

Tea: Simmer 15 to 20 fresh small branchlettes in 2 quarts water for 10 to 15 minutes. At this point, it may be as strong as desired; or the berries may be left in the water to steep and heighten the flavor. Makes a sprightly drink.

Salmon steaks with juniper

Press 5 halved juniper berries into each salmon steak, dot with butter and grill on each side until done. Serve with salt, pepper and lemon wedges.

Manzanita jelly

The little wild berries in canyons near her home in Douglas provide **Ora Sigrist** *the fruit for this jelly.*

1½ quarts manzanita berries
2 quarts water
6 cups sugar
1 bottle liquid pectin or 2 packets pectin
½ cup lemon juice

Stem, pick over and wash the berries. Put in kettle with water and boil rapidly for 10 minutes. Cut down the heat and simmer for 20 minutes or until water is amber. Drip through a colander or jelly bag. Measure 4 cups liquid into saucepan and add 4 cups sugar. Boil for 15 minutes; add 2 cups sugar and the pectin. Boil rapidly for 10 minutes. Add lemon juice and boil for 10 minutes. Pour into sterilized jars. Cool and cover with paraffin.

Using rose hips

Dried rose hips. Dry the berries in slow oven (140 to 150 degrees) and then store in airtight containers. For making tea, coarsely grind dried rose hips and use 1 tablespoon ground rose hips to 1 cup water. Boil until desired strength is obtained and strain. Sweeten to taste. For making jam, cover dried hips in water and simmer until soft. Put through sieve to remove seeds. Use about 1 cup sugar to each 2 cups pulp and 1 teaspoon lemon juice and simmer until of jam consistency.

Fresh rose hips. Cook 4 cups berries with 2½ cups water until tender and sieve out the seeds. Using 1 cup sugar, 2 cups pulp and 1 teaspoon lemon juice, simmer slowly about 20 minutes. Good combined with other fruits such as apple.

Rose hips

Cacti
and
related plants

Prickly pear cactus fruit is becoming more widely enjoyed.

Cacti

For flower lovers, the striking red, pink, golden or white blossoms of cacti are one of the rewards of desert living.

But the blossoms are no more fascinating than the plants themselves, which are perhaps the best known of the succulents, the name given to the numerous plants that are able to store water in their leaves, stems or root systems for use during long dry periods.

Hundreds of species make up the cactus family, from the towering giant saguaro to miniatures that hug the ground. Native to the Americas, their main home lies in the deserts of the Southwest and northern Mexico.

Outwardly, cacti have very different appearances, but their flowers, fruits, seeds, etc., are similar botanically. Some cacti have thorns that not only protect them from animals but offer shade as well.

For centuries, desert people have used the fruits, flowers and seeds of these plants for food and drink, building materials and medicines.

When grown in today's urban settings — in yards or as house plants — cacti sometimes are overwatered or overfertilized. This is regrettable, because the plants prefer the environment to which they have become accustomed since the time of the Eocene jungles.

Cacti share the desert landscape with other "dry plants" *(xenophytes)*, such as the agave, ocotillo and yucca. These differ from succulents in that they do not store water. They survive because they can get by on small amounts. They, too, provided food for desert people. (A few of the myriad of other desert edibles are described briefly in the section on nuts, seeds and desert beans.)

Arizona law protects many native plants from destruction, but the fruit and seeds may be gathered with discretion. Permits to gather plants must be obtained from the Arizona Commission of Agriculture and Horticulture before they may be transplanted.

Agaves

Botanical name: *Agave species*

This remarkable desert succulent, a member of the Amaryllis family, blooms only once and then dies. Since it may grow many years before its blossoms appear, it is known as the "century plant."

The base of the agave, or maguey, is composed of large blades or spears. At blooming time, a single stalk resembling a giant asparagus spear rises 10 feet or more from the center. The succulent is an attractive ornamental in urban yards in elevations of 1,200 to 8,000 feet.

Another name for agave is *mescal*, so called by the Spaniards because the

Mescalero Apaches depended on the plant for food, fiber and medicine. Agaves vary in size from 1 foot to 6 feet in height at maturity. A large species that grows in Mexico is used to make tequila.

The Indians baked the heart of certain types of agaves that had fairly fleshy leaves for food. They also baked the bases of the leaves and mashed them to extract the juice and pulp from which they made a thick, dark brown syrup. The plant provided calcium and minerals to their diets, along with sweetness and calories.

The varieties suitable for cooking in the Arizona portion of the Sonoran Desert are smaller than the large, gray-blue agaves that many homes use as desert ornamentals in their yards. And under no condition should agave be eaten raw, as it is poisonous in that state. Also, gloves and protective clothing should be worn because the raw juice is very irritating to the skin.

Agave chiffon pie

This agave chiffon pie has become a tradition at the annual spring desert-survival food project at Flowing Wells Junior High School, Tucson. The pie recipe was developed by biology teacher **Gwen Curiel** *after students collected the agave hearts under the direction of biology teacher* **David Thomas.** *Thomas says the agave used should be the palmerii or parryi.*

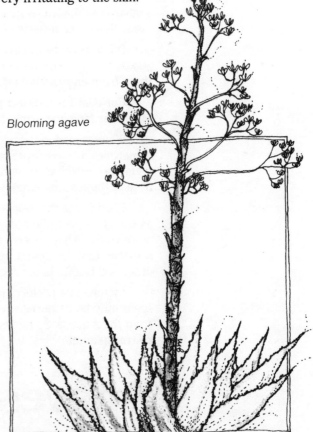

Blooming agave

3 to 4 cups agave juice
2 eggs, separated
1 cup sugar
¾ cup milk
1 tablespoon unflavored gelatin
¼ cup hot water
½ teaspoon salt
2 cups whipped cream or whipped topping
1 teaspoon pumpkin pie spice, nutmeg or
   cinnamon
1 teaspoon vanilla
1 9-inch baked pie shell
¼ cup sliced almonds

Select a 3- to 4-foot agave of the correct species (see above), and remove blades. (Use gloves and protective clothing.) Cut heart in 1-inch pieces. Cook in water to cover in large pot, for 3 to 5 hours. Remove from heat, mash with potato masher until thick and soupy. Press out milky-looking juice through strainer or cheesecloth. Beat yolks until thick. Combine agave juice, ½ cup sugar, milk and yolks in top of double boiler and cook for 10 minutes, stirring constantly. Remove from heat, stir in gelatin softened in hot water and chill until thick. Beat egg whites until stiff, gradually adding remaining sugar and the salt. Fold the whipped cream into the chilled agave mixture; then fold in beaten egg whites, spices and vanilla. Pour in baked pie shell. Top with almonds; chill.

Barrel cacti

Botanical name: *Ferocactus wislizenii and others*

The name for this handsome, ribbed desert succulent (*visnaga*) is derived from its shape. Some barrels have hatpin-like spines, others hooked spines that would make quite respectable fishhooks. In the home garden, barrels are frost-resistant if watering in fall and winter is reduced.

The old idea that travelers lost in the desert could survive on water stored in barrel cacti is only partly true. The *wislizenii*, a common type in urban settings, does contain drinkable liquid. (In other types — some have red blossoms — there is oxalic acid in the liquid that can cause serious reactions, including upset stomach and the inability to walk.)

The yellow fruit of the common barrel cactus forms in a circle on top and is lemon-like in flavor. When peeled with a potato peeler, it can be made into a straw-colored jelly. The flowers and buds are edible and offer nutrition in the form of minerals and carbohydrates. The somewhat crunchy seeds resemble poppy seeds and are high in protein and oil.

Regrettably, some people make candy from barrel cacti by cutting them up and boiling the pulp with sugar, but those who have sampled the sweet say it isn't very good. This procedure destroys these beautiful, slow-growing plants, and this cookbook does not carry the directions for preparing it.

In the old days, the Papagos sometimes gathered the barrel cacti fruit in late winter when other fruit was scarce. They called it "desperation fruit."

Chollas

Botanical name: *Opuntia fulgida; O. bigelovii and others*

The cholla belongs to the same opuntia family as the better-known prickly pear. It has cylindrical branches joined like sausage links.

Some, such as the "jumping" or "teddy-bear" cholla, are covered with pale, sharp stickers and are hazardous to be near. Sections of the branches have a tendency to latch onto any passer-by at the slightest touch. Other chollas are skimpier on stickers, but also should be approached with caution.

Before they open completely, the tender buds can be steamed as a green vegetable of some delicacy. The buds tend to be gelatinous, but drying before steaming minimizes this characteristic. Admirers say the flavor of cholla buds resembles asparagus combined with artichokes. The fruit also can be eaten.

With persistence, it is possible to gather several gallons of buds in a day. They must be dethorned before cooking, a somewhat difficult process. Low in calories, cholla buds are a fair source of vitamin A, high in calcium and other minerals.

Ocotillos

Botanical name: *Fouquieria splendens*

This desert shrub is spindly branched and thorny, with spring blooms of remarkably handsome red flowers at the end of each long branch. The name *ocotillo* is Spanish for "little torch."

The ocotillo is a dry desert plant often used in urban yards as a decorative plant.

During most of the year, its branches, which grow from a central trunk at ground level, appear to be dry sticks. The leaves are green and tiny, and when they fall, reveal thorns. The branches are often cut and stuck in the ground to make fences that may sprout in spring. The branches also are used to shade ramadas or lean-tos.

The flowers, though sparse, can be steeped in cold water to make a pleasant-tasting drink when combined with other juices. The "tea" is slightly pink. The amount to be used is equal parts of water and flowers. The flowers contain sugars and some minerals.

Prickly pears

Botanical name: *Opuntia phaeacantha or O. engelmannii; O. ficus-indica and others*

Perhaps the best known of the edible cacti is the prickly pear or flat opuntia. Its distinctive stems or pads, growing one out of the other, make it easily recognizable.

Some prickly pears are yellowish-green; others have a pinkish or purplish hue. The pads or *nopales* may be 8 inches or more in length. There are also miniature prickly pears. Some varieties have no thorns, but all have sharp barbs or glochids, calling for care in handling.

In spring, the tender new green pads or *nopalitos* can be used raw in salads when silver dollar-size, or peeled, sliced and cooked as a green vegetable when about 5 to 7 inches long. (For those who do not gather their own nopalitos, the vegetable is available at some markets, fresh or canned.)

Prickly pear flowers are usually yellow. The fruits (also called *tunas*) ripen in late summer. The sweet purple-red fruit of the Engelmann prickly pear is popular for eating, but the tunas must be peeled with care, because of their stickers. The flavor of the reddish pulp is pleasant and fruity and imparts a distinctive, deep pink shade to food and drink.

Tunas from the treelike Indian fig or Aztec prickly pear are yellow or red. This fruit also is good to eat. Jelly and other concoctions made with Indian fig tunas is yellowish.

Since prickly pear fruit contains little pectin, both commercial pectin and lemon or lime juice must be added so the juice will jell. The fruit also goes into marmalade or baked goods, and the juice is excellent in drinks and other ways. Both pulp and juice may be frozen.

Nutritionally, prickly pear fruit contains fair amounts of potassium and ascorbic acid and is low in calories (about 40 per 100 grams of edible pulp). The pads contain fair amounts of vitamin A.

A fun procedure for anyone with patience is to make prickly pear wine. There is a somewhat better result when the wine is a blend, according to some experts. They recommend using 25 percent wine from prickly pear and 75 percent from grape, such as Perlette or Thompson Seedless. The grapes add smoothness.

Prickly pear bread

A hearty loaf of whole-wheat and white flour is attractive and flavorful when made with the fruit of the prickly pear, as suggested in this recipe from **Susan A. Knuth,** *Corona de Tucson.*

1½ cups unbleached white flour
1½ cups whole wheat flour
  3 teaspoons baking powder
½ teaspoon salt
⅛ teaspoon mace
¼ cup butter
¼ cup honey
  1 egg
  2 tablespoons grated orange rind
¾ cup prickly pear fruit, peeled, seeded and cut up
¼ cup prickly pear juice
¼ cup orange juice
½ cup milk
½ teaspoon vanilla
  1 cup pecans, chopped

Mix flours, baking powder, salt and mace. In separate bowl, cream butter with honey; beat in egg. Add orange rind, prickly pear fruit and juice, orange juice, milk and vanilla. Add to dry ingredients, stirring only until blended. Fold in nuts. Place in greased 9-by-5-inch loaf pan. Bake at 350 degrees for 1 hour or until done.

Festive prickly pear jellied salad

This jellied fruit salad from Tucsonan **Mary A. Hardy** *is as pretty to see as it is pleasing to the palate. She makes it in a ring mold and garnishes the platter with a ring of Arizona fruits.*

  2 envelopes unflavored gelatin
  1 cup orange juice
1½ cups prickly pear juice
⅓ cup sugar
¼ teaspoon salt
½ cup dry red wine
⅓ cup lemon juice
  2 cups fresh fruit (grapes, diced orange sections, diced apples, diced pears, etc.)

Soften gelatin in ½ cup orange juice. Juice 8 to 10 washed prickly pears in blender with about 1 cup water. Strain seeds, peel and stickers through 3 layers of cheesecloth. Combine remaining orange juice and 1½ cups prickly pear juice with the sugar and salt and heat almost to boiling. Add softened gelatin and stir to dissolve. Add wine and lemon juice. Chill until slightly thickened. Add the 2 cups fruit and pour into large ring mold sprayed with non-stick spray. Chill until set. Unmold on large round platter and garnish with salad greens and more fruit. Serve with sour cream or mayonnaise. Serves 8 to 10.

Cactus cobbler

*When **Jere Hunter** of Green Valley and her husband have visitors, cactus cobbler is likely to be on the menu. She prepares and freezes the fruit so that she can make the dessert when the fruit is not in season.*

Dough:
  1 cup flour
  ½ teaspoon baking powder
    Pinch salt
  ¼ cup margarine
  5 tablespoons milk

Filling:
  4 cups prickly pear fruit, frozen
    or freshly prepared
  1 teaspoon cinnamon
  ½ teaspoon nutmeg
  3 teaspoons tapioca

Sauce:
  ¼ cup margarine
  ¾ cup sugar
  1 teaspoon vanilla
  6 teaspoons lemonade granules
    Water

Mix flour, baking powder and salt. Cut in margarine until mixture is consistency of corn meal. Add milk and mix until dough forms. Turn onto floured board and roll into a large circle. Cut into 4 pie-shaped quarters.

Defrost cactus fruit or prepare fresh by peeling and removing seeds. Spread cut-up fruit over the 4 pastry pieces and sprinkle with cinnamon, nutmeg and tapioca. Roll up and place in a 6-by-10-inch baking dish.

In a small pan, mix together the margarine, sugar and vanilla. Add lemonade granules with enough water to make 1 cup liquid. Heat sauce until margarine melts and sugar is dissolved. Pour over cobbler. Bake at 350 degrees for 1 hour or until crust is brown. Serve warm or cold. Serves 4.

Puréed or sliced prickly pear

To prepare cactus to be pureed or sliced, cover fruit with boiling water for about 1½ minutes, then drain. Holding fruit with tongs, remove peeling and stickers easily with a sharp knife. Cut peeled fruit in half and scoop out seeds with a spoon. Stickers in your fingers can be removed Indian-style by rubbing your fingers through your hair near the scalp.

Cactus flambé

***Agnes Daniels** of Tucson, who, with her husband, **Ross**, makes prickly pear wine, also uses the fruit in various other ways, such as this suggestion.*

  2 tablespoons butter
  2 tablespoons frozen orange concentrate
  3 tablespoons lemon juice
  ½ cup sugar
  2 cups prickly pear fruit, peeled, seeded
    and cut in pieces the size of cherries
  ¼ cup prickly pear juice
  2 ounces cherry-flavored brandy
  2 tablespoons flour or arrowroot
  ½ cup regular brandy

In a flambé pan or skillet, heat butter, orange concentrate, lemon juice, sugar and prickly pear fruit. Combine prickly pear juice, cherry-flavored brandy and arrowroot. Stir into first mixture and let bubble for 2 minutes.

To flambé: Pour ½ cup warmed regular brandy over fruit. Light and shake until it stops burning. Serve on ice cream. Serves 6 to 8.

Prickly pear-pineapple delight

Martha Schuetz of Tucson devised this dessert when all her jelly jars were filled and there was still prickly pear juice to use.

1 graham cracker
  pie crust, chilled
1 envelope unflavored gelatin
  Scant 1 cup sugar
2 cups prickly pear juice
2 to 3 tablespoons lemon juice (to taste)
1 15-ounce can crushed pineapple, well-
  drained
1 cup whipped topping

Stir gelatin and sugar together in bowl. Bring ⅔ cup prickly pear juice to a boil, add gelatin-sugar mixture and stir until dissolved. Add remaining juice, and lemon juice and stir well. Refrigerate until thick but not firm. Blend in whipped topping. Fold in pineapple and pour into crust. Chill at least 3 hours.

Prickly pear sauce

Sauce from the prickly pear fruit makes an interesting topping for ice cream, sherbet, custard or angel food cake.

¼ cup sugar
 1 tablespoon cornstarch
⅛ teaspoon salt
 1 cup prickly pear purée
  Few drops almond extract
 2 tablespoons lemon juice

Combine sugar, cornstarch, salt and purée. Cook, stirring until slightly thickened and clear, about 5 minutes. Stir in almond extract and lemon juice and cook a few minutes longer. Makes 1¼ cups.

Variation: Add 1 9-ounce can crushed pineapple, undrained, with the purée. Increase the cornstarch to 1½ tablespoons.

Prickly pear fruit

Prickly pear cookies

Nancy Weinert developed these cookies in her University of Arizona experimental foods class. They are pink and pretty.

½ cup butter
¾ cup brown sugar
¾ cup sour cream
 1 egg
 1 teaspoon vanilla
1⅓ cups flour
 1 teaspoon baking soda
¼ teaspoon salt
¼ cup prickly pear juice
¼ cup prickly pear, peeled, seeded and
  diced
¼ cup maraschino cherries

Cream butter and sugar. Blend in sour cream, egg and vanilla extract. Mix dry ingredients and blend into sour cream mixture. Add prickly pear juice and fruit and the cherries. Drop from teaspoon onto greased cookie sheet. Bake at 375 degrees for 15 to 18 minutes. Makes 2½ dozen.

Cactus candy

Make recipe twice, once with the Engelmann (purple) prickly pear fruit and once with the Indian fig (yellow-red) prickly pear fruit, for an attractive serving of candy in two colors.

   1 cup prickly pear juice
   2 envelopes plain gelatin
   ½ cup water
   2 cups sugar
   ⅛ teaspoon salt
     Powdered sugar

Place 7 or 8 washed prickly pears in blender with 1 cup water and liquify. Strain through layers of cheesecloth. Measure 1 cup juice. Soften gelatin in ½ cup water. Bring strained purple prickly pear juice, sugar and salt to the boiling point, add gelatin to hot juice and stir until dissolved. Boil slowly for 10 minutes. Pour into 8-inch square pan. Allow to set at least 12 hours. Cut candy into small squares and roll in powdered sugar.

Repeat the same recipe and directions, using 1 cup yellow prickly pear juice instead of the purple prickly pear. Arrange on candy dish in alternating colors.

Prickly pears, range style

Kirk Barnette *of Phoenix shares instructions on how cowboys prepared the fruit, shown to him by his grandfather.*

When the fruit was ready to eat, the cowboys would pick the ripe fruit, place it on a stick and rotate it over the fire to singe off the stickers. Then the men would peel the fruit, cut it in half, scoop out the seeds and eat the pulp with a pinch of salt. "It tasted great that way — still does," says Barnette.

Prickly pear pops

*Tucsonan **Jessie Curley's** children enjoy frozen ice-pops made with prickly pear juice, easy to achieve if you have a blender.*

   2 cups juice from ripe prickly pear fruit
   1 cup water
   2 tablespoons lemon juice concentrate
   ½ cup sugar

Wash the pears and put in blender (stickers, too) with 1 cup water. Blend, then strain the juice through several thicknesses of cheesecloth.

Add lemon-juice concentrate and sugar. Pour mixture into plastic ice-pop makers (or use paper cups and wooden sticks). Freeze.

Cactus-date conserve

Dates, pineapple and pecans or walnuts make this conserve special.

   2 cups peeled and thinly sliced
     prickly pear pulp
 1½ dozen dates, pitted and cut up
   1 orange, juice and grated rind
   2 slices thinly sliced pineapple
   4 teaspoons lemon juice
   ½ cup pineapple juice
 1½ cups sugar
   ⅓ cup broken pecans or walnuts

Prepare cactus fruit as described on Page 28. Combine all ingredients except nuts and cook slowly until thickened and of right consistency. (Use test in marmalade recipe.) About 5 minutes before removing from heat, add nuts. Ladle mixture into sterilized jars and seal.

Prickly pear jelly

Caution: Always make jelly in small batches. It is not likely to jell in large batches.

Recipe I. The basic recipe from the Pima County Cooperative Extension Service calls for this procedure: Wearing gloves and using tongs, gather about 2 quarts of pears, including a few that are not fully ripe. Hold pears under running water and scrub with a brush to knock off small stickers.

Without peeling, slice pears in large pieces and put in large kettle with enough water to barely cover. Boil until tender, about 20 to 25 minutes. Press with potato masher and strain through jelly bag or two thicknesses of cheesecloth. Spines will come off fruit during this process. At this point, juice may be frozen for making jelly later.

To 2½ cups juice, add 1¾-ounce package powdered pectin. Bring to fast boil, stirring constantly. Add 3 tablespoons lemon or lime juice and 3½ cups sugar. Bring to hard boil. Cook for 3 minutes at a rolling boil. Remove from heat, skim and pour into sterilized jelly glasses. Seal at once with ⅛-inch melted paraffin.

Recipe II. This second version, from **Gloria Thomasson** of Tucson, is made with liquid pectin as follows:

Gather and wash 3 quarts fruit (use tongs, gloves and brush, as outlined in extension-service recipe). Place washed fruit in large 6-quart kettle with 3 quarts of water. Cook for 20 minutes, mash with potato masher and cook additional 10 minutes. Strain juice through several thicknesses of cheesecloth. This amount of fruit should be enough to make 8 or more cups of juice.

Measure 4 cups of juice in large kettle, add juice of ½ lemon (if desired) and 3 cups sugar and bring to a boil. Add another 3 cups sugar and cook and stir until mixture reaches a full boil. Add 1 bottle liquid pectin or 2 packets liquid pectin, and continue cooking over high heat for about 15 minutes. Makes 6 glasses of jelly. Repeat with remaining juice, for another 6 glasses of jelly.

Easy-do juicing

Jeanne A. Fischer *of Tucson suggests an easy way to juice prickly pears for jelly.*

Pick the fruit with kitchen tongs. Dump them in a sink full of water and swish the fruit around with the tongs. Then transfer them to a large, heavy kettle. Add about ½ inch of water and put on the lid. Boil the fruit gently until it is softened. After it starts cooking, use a potato masher to make sure each fruit is crushed open. Continue cooking (about an hour) until fruit is tender. Pour fruit and juice into a bag, drain and there's your juice for jelly.

Further tip: Fischer saves old pillow slips and uses them for draining the fruit, then discards them.

Gift jelly

A very popular gift item from the desert is prickly pear jelly. **Melba M. Gallagher** *of Sierra Vista is among the many who use it this way.*

To make never-fail prickly pear jelly boil 4 cups prickly pear juice with 1 package powdered pectin. Add 5 cups sugar and boil until mixture "sheets" from a metal spoon or clings between the tines of a table fork.

Cactus-lemon marmalade

Citrus always adds a desirable touch to recipes using prickly pear fruit.

¼ cup lemon, thinly sliced
1 cup peeled and thinly sliced prickly
    pear pulp
½ cup sugar

Cover lemon slices with water and soak overnight. Prepare fruit for slicing as suggested on Page 28. Add cactus slices and sugar to lemon slices.

Cook marmalade until mixture thickens, giving a satisfactory marmalade test: Pour a small amount of boiling marmalade onto a cold plate and put in the freezer for about 5 minutes. If mixture gels, it should be ready. (Be sure to remove the marmalade from the heat while making the test.)

Or if you have a jelly, candy or deep-fat thermometer, use it to test. Cook marmalade until thermometer registers 9 degrees higher than the boiling point of water. Ladle into sterilized jars and seal.

Saguaros

Botanical name: *Carnegiea gigantea (saguaro);
Lemaireocereus thurberi (organ-pipe)*

The magnificent giant saguaro abounds only in the Sonoran Desert. Slow-growing, it is often seen with branches upraised to the sky. It has long supplied a sweet red fruit that the Papagos and their relatives, the Pimas, harvest in summer. Dried saguaro ribs and other wood are fastened together into long poles and used to knock the fruit to the ground.

About the size of a hen's egg, the fruit traditionally was made into jam or syrup, or into wine for an annual ceremony of thanksgiving, socializing and merrymaking that preceded the planting of beans, squash and corn. Among other uses: Saguaro seeds were ground into flour or used for chicken feed.

The saguaro has wide-ranging, shallow roots that drink up the desert rains, and its accordion-pleated body expands to hold the moisture until needed. At rainy times of the year, a good-sized saguaro can weigh thousands of pounds.

Large, horn-shaped white blossoms, designated the state flower of Arizona, appear on top of the saguaro limbs in spring. The fruit ripens around mid-June.

Though "cactus camps" are still set up each year to harvest and process the fruit, nowadays the Indians are more likely to purchase their jam and syrup at the store.

Describing a saguaro as "slow-growing" means just that. It grows only about a foot every three years, and does not produce fruit until 40 to 50 years old. The plants can live for 200 years, or more, and can grow to 40 feet in height. The trunks and branches often contain holes made by woodpeckers, and later these holes become homes for other creatures. Many saguaros are lost to bacterial necrosis disease, a form of rotting from old age.

As Arizona's population has grown, it has become necessary to prevent destruction of these magnificent cacti. Federal protection comes in both Saguaro

National Monuments, one east and one west of Tucson, and state laws protect the plant from vandalism elsewhere. A permit is required before transplanting to urban landscapes.

The melonlike pulp around the seeds of the saguaro fruit has a flavor described as "a cross between watermelon and fresh figs."

Kin to the columnar saguaro is the **organ pipe cactus,** whose fruit also is tasty. The organ-pipe has multiple trunks in place of the single trunk of saguaro, and is found in Arizona mainly in Organ Pipe Cactus National Monument, south of Ajo.

The fruit of columnar cacti is rich in vitamins and minerals; its seeds are especially high in protein and oil and can be ground into a peanut-butterlike spread.

Saguaro chiffon pie

Sophie Burden *of Wickenburg makes the most beautiful and delicious of chiffon pies with juice from saguaro fruit during the few days each summer when the fruit is ripe. She uses a long pole with a wire loop on the end to harvest the fruit. No added sugar is required.*

    1 9-inch baked pie shell with
       high, fluted edge
    ⅓ cup saguaro fruit juice
    1 envelope unflavored gelatin
    3 eggs, separated
    ½ cup saguaro fruit juice
       Whipped cream

To extract juice, gather and wash fruit and cut in several pieces with knife or scissors. Drip through jelly bag or old sheet, mashing fruit from time to time with a potato masher.

Soften gelatin in ⅓ cup saguaro juice. Place egg yolks in top of double boiler and stir in ½ cup saguaro juice. Cook and stir over boiling water for 5 minutes or until thick. Stir in softened gelatin mixture and stir for 1 minute more to dissolve gelatin. Remove pan from heat. Beat egg whites until stiff and gently fold into hot mixture. Pour into baked pie shell and chill until set. Whip cream and spread on pie.

Saguaros

Saguaro sundae

For a truly glamorous desert dessert, serve this idea from Tucsonan **Stephanie Daniel.**

Cut fruit in half, peel and with a spoon scoop pulp and its seeds from peeling. Freeze in pint containers. (Seeds are edible — just a little crunchier than strawberry seeds.) At serving time, defrost a pint of the fruit, stir in about ⅓ cup orange liqueur and 1 tablespoon vanilla. Serve over vanilla ice cream to 4.

Saguaro jam

The jam is made without added sugar as the fruit contains sufficient natural sweetening.

Peel ripe fruit of saguaro cactus, using gloves to protect the hands. Soak fruit for about 1 hour. To make a quart of jam you will need about 2½ pounds of fruit.

Drain some of liquid from fruit, leaving it only half covered with water. Boil for ½ hour. Strain off liquid, reserving the pulp. Boil the liquid slowly to a syrup, stirring constantly so it won't burn. Crush pulp and put through sieve to remove the seeds. Add the pulp to the thickened syrup and cook to consistency of jam.

Saguaro jelly

This jelly is a rich red color.

3¼ cups prepared saguaro juice
 1 package powdered pectin
 ¼ cup lemon juice
4½ cups sugar

To prepare saguaro juice, gather the ripe fruit, cut in half and remove the seedy pulp. (The dried fruit lying on the ground may be used, as well.) Place fruit in large kettle and add water to about 1 inch above the fruit. Simmer until tender and strain through a fine colander or bag. Pour the juice into a large kettle and add the powdered pectin and the lemon juice. Stir until mixture reaches a full boil; add the sugar. Bring to a boil, stirring all the while. Boil hard for 1 minute. Pour into hot, clean glasses and top with paraffin. Makes 3 pints.

Organ pipe jam

*Tucsonan **Jane Eppinga** supplies these directions for using the fruit of the organ pipe cactus (pitahaya), as the women in Baja California do it.*

3 cups prepared ripe pitahaya fruit
2 cups chopped, pitted dates
½ glass fresh orange juice
 Juice and rind of 1 lime
1 cup panocha or dark brown sugar
½ cup chopped nogales (native walnuts)

Cut fruit in half and scoop pulp from skin with a spoon. Remove the seeds. Combine all ingredients except nuts and cook and stir until mixture is thick (about 15 minutes over moderate heat). Add nuts and cook for 5 minutes. Pour into sterile jars and seal with paraffin.

Other saguaro uses

*Tucsonan **Karen Reichhardt** provides these instructions for using saguaro fruit.*

Saguaro syrup. Delicious over ice cream or pancakes, the reddish saguaro syrup is easily made by boiling down the strained juice until it is of syrup consistency. No sugar needs to be added because of the sweetness of the natural juice.

Saguaro seeds. After straining the juice from the fruit, the seeds are washed to remove any bits of pulp that clings to them, and then spread to dry. The seeds may be ground in a grain grinder into a nutmeal and cooked with water as for oatmeal, adding sugar and salt to taste. Because of the highly nutritious quality of the seeds, the Indians call this dish "breakfast of champions."

Dried saguaro fruit

The fruit — pulp and seeds — when dried is a candylike sweet that is very good to eat, says **Mahina Drees** *of Tucson, who has the handsome plants in her yard.*

Gather the fruit as it ripens. (While saguaro fruit has fewer stickers than prickly pears, some people may wish to wear gloves.) Cut fruit in half and scoop pulp and seeds out of the skin with a teaspoon. The seeds are somewhat like poppy seeds only a little larger.

Spread the fruit halves to dry on wax paper. Sun drying takes about three days. The fruit turns very dark as it dries.

Roll the dried fruit into balls and store in a plastic bag in the freezer. Serve for a snack.

Yuccas

Botanical names: *Yucca baccata (wide-leaf); Yucca elata (narrow-leaf)*

Yuccas are drought-resistant desert dwellers that often are grown in urban settings. Edibles come from the banana yucca (also called datil), a wide-leaf yucca, and the soap tree (also called palmilla), a narrow-leaf yucca.

The banana yucca is stemless and has small clusters of flowers that produce reddish, banana-shaped fruit. This can be eaten raw or cooked, sweetened and used as any stewed fruit. Native Americans also ground the seeds into meal. The plant grows throughout the Southwest at elevations from 3,000 to 6,000 feet, often along with piñon and juniper.

The palmilla grows as high as 12 feet. Its stem is hidden with long, straw-colored, grasslike leaves with green leaves at the top. In hot weather, its distinctive flower stalks rise as much as 6 feet above the rest of the plant and bear fragrant, snowy blossoms. This yucca grows at elevations from 1,500 feet to 6,000 feet.

Native Americans baked the stems and cooked and ate the blossoms of the palmilla (after discarding the bitter centers). Its fiber was made into ropes, sandals and cloth. The root furnished a detergent-like substance that could be pounded in water to make suds for washing clothes and shampooing hair. There were medicinal uses for yucca as well.

Mexicans also used yucca plants, and the flowers were a common item for sale in the food markets of Mexico well past World War I.

Yucca blossoms contain carbohydrates and some minerals; the fruits have a good complement of vitamins and minerals; and the stalks have some protein and vitamins.

Uses for yucca

In salad: Separate Yucca elata petals and soak in water. Drain and use in tossed salad. Flavor resembles Belgian endive.

In soup: Sauté washed Yucca elata petals in butter and add to cream of chicken soup, with a dash of dill weed.

Yucca omelet

This recipe is based on a suggestion in a 100-year-old book of household hints. The petals of the blossoms, but not the bitter centers, are used.

1 cup Yucca elata petals
6 eggs
    Butter

Plunge washed petals in boiling water for a few minutes and then drain. Make individual omelets by beating 2 egg yolks at a time and 2 whites at a time until whites stand in soft peaks but are not stiff. Fold yolks and whites together for each omelet. Add ⅓ of the petals and pour into well-buttered hot skillet. Brown lightly over medium heat, turn and brown on other side. Fold in half and serve topped with butter. Serves 3.

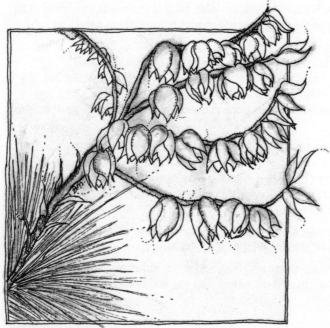

Yucca blossoms

Citrus

Oranges are the best-loved of all the citrus fruits.

Citrus

For many desert dwellers, the rewards of faithful watering, pruning and fertilizing are not only citrus fruits for many months but trees with bright greenery for year-round beauty. Virtually every home where climate allows — especially in Maricopa and Yuma counties — boasts one or more varieties. Some portions of Pima County are satisfactory for citrus.

Citrus fruits have Asiatic origins. They are mentioned in early Chinese writings, and as travelers and traders carried them westward to Europe and North Africa, new admirers learned of their eating delights.

The first citrus seeds reached the New World in 1493 during the second voyage of Columbus to Haiti. Mexico had citrus by 1518, and by 1565 the trees were started in Florida. Citrus fruits were planted in Arizona early in the 18th century at missions established by Father Eusebio Kino.

The fruits have been valued for centuries for their health-giving properties. Modern vitamin and mineral analysis shows why: All citrus fruits contain a considerable amount of vitamin C, plus a fair amount of vitamin A, B vitamins and minerals, especially potassium. They are low in sodium.

In Pima County (Tucson), it is too cold in the river bottoms for citrus, but on the south and west slopes of the Catalina and Rincon and Tucson Mountain foothills, citrus-growing is generally successful. These slopes are microclimate areas that receive and store sunshine all day and radiate the warmth at night.

In the second- or middle-elevation areas of Tucson, success is marginal. But citrus trees can produce fruit in this zone, especially if planted on south walls and in sheltered locations. In this marginal area, early ripening, juice oranges are preferable, because later ripening ones may receive frost damage. Lemons and limes do not give dependable results.

Arizona raises commercial crops of oranges, lemons and grapefruit on more than 56,000 acres. Much of the fruit is sold through the marketing cooperative Sunkist Growers.

When should the home grower pick his citrus crop? Except for mandarins, which tend to dry out, citrus should be taken from the tree only as needed. The fruit is ready when it tastes good, (sample from time to time) and it keeps for many weeks on the tree. During the desert's rare freezing weather, the fruit can be picked and the juice frozen.

Grapefruit

Botanical name: *Citrus paradisi*

Grapefruit trees thrive happily in our hot desert climate. If their bright yellow fruit is left on the tree long enough to develop some sweetness (this requires several months), the trees display the blossoms for the next crop along with the fruit.

Grapefruit is easy to grow. It ripens November to June, and any home orchard with a tree or two usually has plenty of extras for gifts. Popular home varieties are Redblush, Ruby Red and Marsh.

Where grapefruit got its name is not known. Some food historians think it was derived from the fact that the fruit grows in clusters, giving it a far-fetched resemblance to clusters of grapes. The fruit is white, pink or red, and the deeper the color, the higher the vitamin A content.

Grapefruit and avocado salad

Tucsonan **Sue Scheff** *likes to make her citrus-avocado salad, which can be served with any kind of Mexican food.*

2 grapefruit, sectioned
  (or use 4 oranges)
1 bunch green onions, chopped
1 avocado, peeled and sliced
1 head lettuce, shredded
4 tablespoons oil
2 tablespoons vinegar
  Pinch of salt, sugar, dry mustard
  and pepper

Combine oil and vinegar with salt, sugar, mustard and pepper, and toss with other ingredients.

Grapefruit combo

Grapefruit ice is the topping for melon balls, creating an unusual appetizer course.

1 cup grapefruit juice
2 tablespoons sugar
4 cups honeydew, or other melon balls
4 sprigs mint (optional)

In a 9-inch cake pan, mix grapefruit juice and sugar until sugar dissolves; freeze for 2 hours or until solid. Using a metal spoon, scrape ice into shavings. Put back into freezer until serving time. To serve: Put melon balls in individual serving dishes or stemmed glasses, and spoon about ½ cup shaved grapefruit ice on top of each. Garnish with mint, if desired. Makes 4 servings.

Grapefruit three-bean salad

When you've run out of ideas for using all your grapefruit crop (and so have your neighbors), try this unusual salad, a variation of the popular three-bean salad.

2 tablespoons butter or margarine
½ cup chopped onion
1 tablespoon cornstarch
1 cup grapefruit juice
2 tablespoons cider vinegar
1 tablespoon soy sauce
¼ cup packed brown sugar
2 cups diagonally sliced celery
1 1-pound-4-ounce can kidney beans,
  drained
1 1-pound-4-ounce can chick peas, drained
1 1-pound can cut green beans, drained
2 cups grapefruit sections

In large skillet, melt butter. Add onion and cook until tender. In medium bowl, mix cornstarch and grapefruit juice, add to skillet with cider vinegar, soy sauce and brown sugar and mix well. Bring to a boil and cook until sauce is thickened, stirring constantly. Stir in celery, drained kidney beans, chick peas and green beans. Cook over low heat for 10 minutes. Add grapefruit sections and heat. Serves 6.

Arizona country- captain chicken

Sue Cosor of Chandler finds this chicken and citrus combination "out of this world."

3 pounds broiler-fryer chicken,
   cut up, skin removed
1/3 cup flour
1 tablespoon salt
1/4 teaspoon pepper
1/4 cup shortening
1 cup each chopped onion, green pepper
1 garlic clove, minced
1 1/2 cups water
1 14-ounce bottle ketchup
2 teaspoons curry powder
1/2 teaspoon ground thyme
1 grapefruit and 1 orange, sectioned
1/2 cup raisins
   Slivered almonds

Coat chicken with a mixture of the flour, salt and pepper and brown in hot shortening. Remove from pan. Sauté onion, green pepper and garlic in drippings. Add water, ketchup, curry powder and thyme, return chicken to pan and turn to coat each piece with the sauce. Bake, covered at 350 degrees for 45 minutes, turning occasionally. Top with the grapefruit, orange, raisins and slivered almonds. Bake another 20 minutes or until chicken is done. Serve with rice. Serves 6.

Grapefruit vinaigrette

For brightening a fruit salad or tossed greens, this quick dressing with grapefruit juice is a pleasing change.

1 cup vegetable oil
3/4 cup grapefruit juice
1/4 cup lemon juice
1 tablespoon sugar
1 teaspoon salt
1 teaspoon paprika
1 teaspoon chopped onion
1/8 teaspoon ground pepper
1 clove garlic, cracked

Put all ingredients, except garlic, in blender. Cover and blend for 15 seconds until smooth. Add garlic and chill for 1 hour before serving. Makes 2 cups dressing. Good served over fresh fruit salad or tossed greens.

Grapefruit cups

Serve this suggestion as an appetizer or with the meat course.

2 grapefruit
1/4 cup butter or margarine
1/3 cup orange liqueur
2 teaspoons sugar
1 large apple, cored, peeled and shredded
1/3 cup seedless grapes, halved
1/3 cup raisins
1/3 cup chopped nuts

To prepare grapefruit shells with saw-toothed edge, mark a line in the center around the grapefruit. Insert pointed knife into center of grapefruit at an angle to make one side of a point; remove knife; insert to make opposite side of point. Continue around grapefruit, following line. Pull halves apart. Cut around each section, loosening fruit from membrane. Reserve sections.

Melt butter in saucepan; add orange liqueur and sugar. Stir until sugar dissolves. Add shredded apple, grapes, raisins, nuts and reserved grapefruit sections. Spoon into grapefruit shells. Bake at 350 degrees for 10 minutes. Serve hot with meat or poultry or as a first course. Serves 4.

Grapefruit ambrosia

This ambrosia is served warm for an appealing change.

1½ cups grapefruit sections
1½ cups orange sections
3 tablespoons shredded coconut
2 tablespoons honey

Combine ingredients in skillet and mix gently. Warm over low heat for 3 to 5 minutes. Serves 3.

Arizona grapefruit-sour cream pie

This recipe was developed by mistake one time by **Juanita McLennan** *of San Manuel when she thawed frozen grapefruit instead of the frozen crushed pineapple she had intended to use for her pineapple cream pie. It turned out great.*

1¼ cups sugar
⅓ cup flour
Dash salt
Sections and juice of 2 large grapefruit
2 large or 3 medium eggs, separated
¾ cup sour cream
1 teasooon orange extract
1 tablespoon margarine
1 9-inch baked pie shell

In large saucepan combine sugar, flour and salt. Stir in grapefruit and juice and cook until mixture thickens, stirring constantly. Add slightly beaten egg yolks to hot mixture, cook for 2 minutes while stirring. Remove from heat and stir in sour cream, orange extract and margarine.

Pour into baked pie shell. Top with coconut meringue and bake at 375 degrees for 20 minutes.

Coconut meringue: Beat egg whites until stiff and add ¾ cup sugar and 1 teaspoon orange extract. Beat until stiff peaks form. Gradually fold in ¾ cup flaked coconut. Sspread on pie and bake at 375 degrees for 20 minutes.

Grapefruit pie

Tucsonan **Emogene Uhl's** *grapefruit pie is another good way to use fresh grapefruit from the home orchard.*

6 tablespoons cornstarch
1¼ cups sugar
¼ teaspoon salt
½ cup cold water
2 cups grapefruit juice
3 eggs, separated
1 teaspoon grated grapefruit rind
1 tablespoon butter or margarine
1 baked 9-inch pie shell

Combine cornstarch, sugar, salt and cold water and stir until well mixed. Then add the grapefruit juice. Cook over medium heat, stirring, until mixture comes to a boil. Continue cooking for 5 minutes, stirring constantly. Remove from heat. Add beaten egg yolks and rind, slowly stirring. Then add butter and stir until mixed well. Let cool for 10 minutes. Pour in baked pie shell.

Top with meringue, made this way: Beat the 3 egg whites with ¼ teaspoon salt until frothy; beat in 6 tablespoons sugar and continue beating until meringue is stiff and glossy. Swirl meringue lightly on pie, sealing to edge of crust. Bake at 375 degrees for 10 minutes or until brown.

Grapefruit with lemon ice

Grapefruit sections flavored with vodka and topped with lemon ice make an attractive dessert.

2 cups plus 2 tablespoons sugar
4 cups water
  Grated rind of 2 lemons
2½ cups fresh lemon juice
1 large, juicy grapefruit
¼ cup vodka

Combine 2 cups sugar with the water in a saucepan and boil for 5 minutes. Add the lemon rind and lemon juice and cool. Chill thoroughly. Pour the mixture into the container of an electric or hand-cranked ice cream freezer, then freeze and pack according to the manufacturer's instructions.

Peel grapefruit and section it. Put the sections in a mixing bowl and add the remaining 2 tablespoons of sugar and the vodka. Chill until ready to serve. Spoon out individual portions of lemon ice into stem glasses or dessert cups and garnish with grapefruit sections and the vodka-sugar syrup. Keep leftover lemon ice in the freezer. Serves 6 to 8.

Grapefruit cake

*Tucsonan **Polly Swineford's** cake has a piquant flavor from the fresh grapefruit juice and peel in the batter.*

2 cups sifted cake flour
1 cup sugar
2 teaspoons baking powder
  Pinch salt
5 eggs, separated
¼ cup fresh grapefruit juice
⅓ cup water
⅓ cup salad oil
1 tablespoon grated grapefruit peel
¼ teaspoon cream of tartar

Sift dry ingredients. Combine egg yolks, grapefruit juice, water and oil in a small bowl. Add dry ingredients. Stir in grated peel. Beat egg whites until foamy. Add cream of tartar and beat until stiff but not dry. Fold egg yolk mixture into whites (just to blend). Bake at 350 degrees in 2 9-inch wax paper-lined cake pans for 25 to 30 minutes. Cool and frost.

Baked grapefruit halves with honey

Nuts, raisins and a little honey sweeten this baked grapefruit dessert.

1 tablespoon butter or margarine
1 tablespoon honey
¼ teaspoon cinnamon
½ cup raisins
½ cup chopped nuts
¼ cup shredded coconut
2 grapefruit

Melt butter in medium saucepan, remove from heat and stir in honey, cinnamon, raisins, nuts and coconut. Cut grapefruit in half; remove core. Cut around each section, loosening fruit from membrane. Spoon raisin filling into the middle of each grapefruit half and spread on top. Place grapefruit in shallow baking pan. Bake at 350 degrees for 15 minutes or until hot. Serves 4.

Rosy grapefruit-sour orange marmalade

Combining grapefruit with sour oranges results in a very pleasing marmalade, Tucsonan **Gloria Thomasson** *has found. The grenadine gives it a rosy hue.*

6 sour oranges
1 grapefruit
  Water
¼ teaspoon baking soda
⅓ cup grenadine syrup
1 package powdered pectin
6 cups sugar

Wash oranges and grapefruit. Cut grapefruit in quarters, remove seeds and chop fruit in 1-inch chunks. Add 1 cup water to blender jar along with grapefruit, and grind. Pour into 1 gallon jar or large 4-quart mixing bowl.

Cut sour oranges in quarters. With sharp knife, cut out center white membrane and seeds. Grind in blender with 1 cup water. Pour into container with grapefruit. Add 2 quarts water, the baking soda and the grenadine. Let stand overnight.

Next day, pour mixture into 6-quart covered kettle. Boil, covered, for about 20 minutes. Remove cover and set aside to cool.

An hour later, later the same day or the next day, measure 6 cups prepared fruit into a 6- to 8-quart kettle. Stir in powdered pectin, and cook over high heat until mixture comes to a hard boil. Add sugar all at once. Continue cooking over high heat, stirring constantly until mixture comes to a hard boil that cannot be stirred down. Boil hard 1 minute. Remove from heat, skim, pour into sterilized jars. Cover with ⅛ inch paraffin or seal. Makes 7 or 8 glasses.

Repeat with other 6 cups prepared fruit.

Note: Thomasson said she found it fit her busy schedule to spread out the preparation time, as suggested here. Also, she said, the holding time seemed to make the flavor less bitter.

Grapefruit jelly

Louise Bonney, *who shares her time between Oregon and Tucson, likes to make grapefruit jelly when she's in Arizona.*

4 cups grapefruit juice
5 cups sugar
1 box powdered fruit pectin

Measure juice in large pan (not aluminum). Mix in pectin and bring to a boil. Stir in sugar all at once. Bring to rolling boil and boil for 3 minutes. Remove from heat, skim and pour into sterilized jelly glasses. Seal.

Grapefruit blossoms

Non-alcoholic grapefruit punch

Grapefruit and tangerine juice combine to spice this party punch.

8 cups grapefruit juice
1 pint ginger ale
2 cups tangerine juice
Lime slices, mint

Chill all ingredients thoroughly. Combine citrus juices with ginger ale in punch bowl. Add cracked ice. Garnish with lime slices and mint. Makes 24 ½-cup servings.

Citrus punch

Grapefruit juice adds its own special flavor, along with other citrus juices in this punch recipe from **Terry Mikel** *of Yuma.*

2 cups sugar
4 cups water
½ cup lemon juice
½ cup lime juice
4½ cups orange juice
2 cups grapefruit juice
2 cups pineapple juice
1 750 milliliter bottle medium-sweet white wine
1 750 milliliter bottle champagne

Boil for 1 minute the sugar, 2 cups water and lemon juice. Stir in remaining water and cool. Stir in lime juice, orange juice and grapefruit juice and pour over ice. Add pineapple juice, white wine and champagne. Garnish with citrus slices, strawberries, mint, etc.

Kumquats

Botanical name: *Fortunella margarita*

Close kin to the citrus, but hardier, kumquats are miniature, elongated, orangelike fruit mostly used for marmalade. Among other uses, a half dozen or so of the mildly acid fruit give flavor and tenderness to meat or poultry dishes. The juice adds a nice touch to beverages, in place of limes or lemons.

Kumquats also can be eaten out of hand, the rind, too. It is only slightly bitter.

In China, the fruit is known as *Chin kan* or "golden orange," and is used more often in cooking than in this country.

Kumquats are ideal ornamentals for urban yards, because the miniature fruit stays on the tree year-round, along with the scented blossoms of the oncoming crop.

The kumquat crosses well with other citrus, each as decorative as the other. Among the hybrids are the **limequat, orangequat** and **citrangequat**, all developed by the U.S. Department of Agriculture.

Another hybrid is the **calamondin**, a natural cross between the kumquat and the mandarin (tangerine). The calamondin is tiny, round and very tart. Though it can be eaten out of hand, the tender peeling along with the pulp, it is usually cooked like a kumquat.

For home growing, successful kumquat varieties are Nagami and Meiwa.

Calamondin chicken

A handful of calamondin limes makes a distinctive sauce for a top-of-the-stove chicken casserole.

8 to 10 calamondin limes
½ cup water
2 tablespoons brown sugar
¼ teaspoon ground cloves
½ teaspoon salt
1 broiler-fryer chicken, cut up
Oil

Wash and halve calamondin limes and remove seeds. Combine brown sugar and water in saucepan and stir in calamondin halves, cloves and salt. Simmer, covered, for about 10 minutes. Brown chicken pieces on all sides in oil. Pour sauce over chicken, cover and simmer until chicken is tender, turning pieces once or twice in sauce and adding small amount of water, if required. Serve topped with sauce. Serves 4.

Kumquat preserves

This recipe is from **Hanna Lundberg** *of Tucson, a many-times ribbon winner at the Pima County Fair.*

Work with 4 cups kumquats at a time. Clean fruit and puncture deeply at one end. Let stand in covered salted water overnight, using 1 teaspoon salt to 1 quart water. Drain.

Drop fruit into boiling water to cover. Cook until tender. Boil 2 cups sugar and 1 cup water to make a syrup. Add drained fruits and cook until clear and transparent. Leave fruit in syrup two or three days for good plumpness, shaking kettle or turning fruit occasionally.

Reheat fruit, pack in sterilized jars. If a thicker syrup is desired, add 1 cup sugar to syrup when fruit is reheated. Makes about 5 ½-pint jars.

Note: Follow the same method when preserving calamondin limes.

Stuffed kumquats

Kumquats, candied and stuffed with nuts, are an attractive sweet to serve along with other bonbons.

Wash 4 cups whole kumquats (about 1 pound), cover with water and boil for 5 minutes. Drain. Cut in half lengthwise. Combine 2 cups sugar and 1 cup water and bring to a boil. Drop in kumquat halves and cook for 10 minutes. Cover and let stand overnight. Cook for 20 minutes longer the next day and lift out of syrup. Place on wax paper to cool. Fill each kumquat half with a toasted pecan half and roll in granulated sugar.

Spiced kumquat preserves

In this version, stick cinnamon and lemon slices add their flavor.

2 cups kumquats
1 cup water
2 cups sugar
1 stick cinnamon
1 lemon, sliced thin

Wash and drain fruit. Cut a small gash crosswise in each kumquat. Cover with water and bring to a boil. Cook for 5 minutes and drain. Make a syrup of sugar and water and add cinnamon and lemon. Drop fruit into syrup and boil for 10 minutes. Cover and let stand overnight. Cook again, uncovered, for 10 minutes and let stand overnight. Bring to a boil again and cook until fruit is clear and syrup thick. Pack in clean jars while hot. Cover with hot syrup and seal. Makes about 4 ½-pints.

Note: Recipe may be used for calamondin limes as well.

Kumquat marmalade

Kumquats and lemon make flavorful marmalade in this recipe from **Leora Patterson** *of Tucson.*

    2 cups thinly sliced firm
        kumquats (discard seeds)
    1½ cups thinly sliced lemon
    6 cups water
        Sugar

Combine all ingredients except sugar, and let stand overnight. On the second day, bring to a full boil and cook for 20 minutes, stirring constantly. Again let stand overnight. On the third day, measure the fruit mixture and measure 1 scant cup of sugar for each cup of fruit. Bring fruit to a boil before adding sugar. Boil gently for 20 minutes or until it jells when dropped from a spoon. Pack in jars and seal or use paraffin to seal. Makes about 8 ½-pints.

Calamondin lime marmalade

Marilla Bixler *of Tucson likes to make this marmalade without added lemon, to emphasize the superb flavor of the calamondin lime.*

    3 cups prepared calamondin limes
    4½ cups water
        Sugar

Wash, slice finely and seed fruit. Put in kettle with water and bring to a rolling boil. Boil for 30 minutes. Let stand for 24 hours. Measure cooked mixture and add 1½ cups sugar to each 1 cup fruit. Bring to rolling boil. Boil for 15 minutes or until marmalade spins a thread. Place in small jars and cover with paraffin. Makes 8 ½-pints.

Kumquats and sweet potatoes

Steam 2 or 3 sweet potatoes until tender, peel and slice. Slice 6 to 8 kumquats, remove seeds and steam in small amount of water until tender.

Layer sweet potatoes, kumquat slices, brown sugar and butter in casserole; top with buttered crumbs. Bake at 350 degrees until heated through. Serves 4.

Lemons

Botanical name: *Citrus limon*

Lemonade and lemon pie. These are two favorites that are always better made with fresh lemons. The fruit also brightens many a salad or fish dish, or substitutes for salt in low-sodium diets.

Lemon trees do best in the low desert, but can be grown in the Tucson area in protected areas. For home growing, Lisbon is a popular choice. Ponderosa, another choice, is a lemon-citron hybrid.

California began growing lemons commercially to provide an antidote for scurvy, a vitamin C deficiency disease that affected the miners during the gold rush of 1849. One lemon will provide 80 to 90 percent of an adult's daily need for vitamin C.

Lemon orchards are now a year-round big business, not only in California but in Arizona as well. The commercial lemons in the winter come mostly from desert areas in the two states, and during the summer from coastal areas in

California. Lemons are picked for sale as soon as they achieve desired size and juiciness, even though they may still have green skins. They are marketed when they turn yellow.

Home-grown lemons, however, can be left on the tree until the skin is sunny gold. If unexpected cold weather might cause losses, the lemons can be juiced and frozen in ice-cube trays, then transferred to plastic bags for storing.

Storing lemons

Lemons should be stored in the refrigerator. Leaving them at room temperature for an hour or more or warming them in water before using will facilitate the flow of juice.

Before juicing a lemon, roll it on a counter top, exerting pressure with the palm of your hand to crush the membranes lightly. This will also increase the juice yield. An average lemon contains about 3 tablespoons of strained juice.

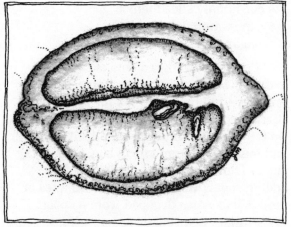

Lemon

Peruvian ceviche

Juan Reynoso brought the directions for making this popular South American dish with him when he moved to Tucson years ago.

2 pounds cabrilla (or other
   firm-fleshed ocean fish)
2 large white onions
1 clove garlic
3 to 5 medium jalapeños
2 cups lemon juice (about)
   Salt

Cut fish into bite-size pieces; wash well and put in large bowl. Pour lemon juice over fish to cover (about 2 cups). Slice onions in a separate bowl and stir with 2 teaspoons salt, 2 cups water and 2 tablespoons lemon juice. Drain well. Add onion to fish. Mash garlic clove and add. Remove all seeds from peppers, slice and add to fish mixture. Mix well, add salt to taste, cover and let marinate at room temperature 1½ to 2 hours. Makes appetizers for 8 or more.

Lemon-berry bread

*Grated lemon peel adds aroma and flavor to this bread from Phoenician **Nancy Kendt**.*

½ cup shortening
1 cup sugar
2 eggs, slightly beaten
1¾ cups flour
1 teaspoon baking powder
½ teaspoon salt
½ cup milk
½ cup chopped pecans
   Grated peel of 1 lemon
½ cup blueberries

Cream shortening and sugar. Add eggs. Sift flour with baking powder and salt and add to first mixture, alternately with the milk. Stir in nuts and lemon peel and blend well. Fold in rinsed and drained blueberries. Bake at 350 degrees in 5-by-9-inch greased and floured pan for about 1 hour. While bread is still hot, top with glaze. Cool and remove bread from pan.

Glaze: Combine juice of 1 lemon with ¼ cup sugar.

Lemon bread

*Making lemon bread was a rare and expensive treat until the family moved to Arizona. Now they have it often, says **Anne Lopez** of Tucson. It has often been a prize-winner.*

6 tablespoons softened butter
1 cup sugar
2 beaten eggs
2 lemons, peeling grated
1½ cups unbleached flour
1 teaspoon baking powder
½ teaspoon salt
½ cup milk
½ to ¾ cups coarsely chopped,
 lightly toasted pecans
 Glaze

Cream butter and sugar. Add eggs and lemon peel. Sift flour, baking powder and salt together. Add alternately with milk. Fold in nuts. Pour into greased and floured loaf pan. Do not use a glass pan. Bake at 350 degrees for 1 hour or until done and lightly browned on top. Remove from oven and let cool in pan for 15 minutes. Dissolve sugar in lemon juice over low heat. Pour over bread and let stand for 10 minutes. Bread will absorb most of the glaze that is on top and sides of the bread.

Glaze: Combine juice of 2 lemons and ⅔ cups sugar.

Lemon-yogurt dip

For a delightful fruit dip to serve with melon spears, strawberries, etc., try this suggestion.

1 cup cottage cheese
1 8-ounce carton lemon yogurt
⅓ cup chopped salted nuts
2 teaspoons grated lemon peel
 Strawberries, melon spears, orange
 slices, etc.

Beat cottage cheese in small mixing bowl on high speed of mixer until almost smooth, about 5 minutes. Stir in yogurt, nuts and lemon peel. Chill, covered, for 1 to 2 hours to allow flavors to blend. Serve with fruit as dippers. Makes about 2 cups.

Spicy lemon dressing

This well-spiced dressing gives the right touch to a cold cooked vegetable salad.

½ cup vegetable oil
¼ cup fresh lemon juice
1 tablespoon soy sauce
½ teaspoon sugar
½ teaspoon ground ginger
1 clove garlic, cut in half
 Dash salt and pepper

Combine ingredients and use on cooked asparagus or broccoli spears with onion-ring garnish, or other vegetables combined with onions. Makes about ¾ cup dressing.

Poppy seed lemon dressing

*A tasty lemon dressing that is great for someone on a low-sodium diet comes from **Sharon Sass** of Tucson.*

½ cup lemon juice
¼ cup honey
2 tablespoons oil
1 teaspoon poppy seed

Combine ingredients and serve with fruit or tossed salad. Makes ¾ cup. The sodium count is 15 milligrams per tablespoon.

Fruit salad dressing

This recipe is a special one for serving to company from Tucsonan **Nadine Oftedahl's** *mother, who was known for her good German cooking back in Illinois.*

½ cup sugar
3 tablespoons flour
½ teaspoon salt
2 eggs, beaten
¾ to 1 cup pineapple juice
2 tablespoons lemon juice (or vinegar)
2 tablespoons butter
½ cup whipping cream, whipped

Mix sugar, flour and salt. Beat eggs well; add pineapple juice, then lemon juice. Add sugar, flour and salt mixture. Cook to a smooth dressing, stirring constantly. Remove from heat; add butter and stir. Whip cream and when first mixture is cool, combine with the whipped cream. Makes 1 quart. Good with any combination of fruits.

Lemon-pineapple dressing

Tucsonan **Winogene W. Earl** *sent us a dressing recipe that was her mother's and has been used by others in her family for many, many years on fruits, salads, puddings, ice cream, etc.*

1 cup juice from canned pineapple
    slices or chunks
    Juice of 2 lemons
2 eggs, well-beaten
½ cup sugar
1 tablespoon flour

Bring pineapple juice and lemon juice to a boil. Cool a bit and stir into 2 well-beaten eggs. Combine sugar with flour and add to mixture. Return to stove. Let come to bubbling boil on low heat, stirring constantly. Let bubble 2 or 3 minutes. Cool, bottle and refrigerate. Will keep up to 10 days in the refrigerator. Use with or without equal parts of whipped cream on fruit, puddings, ice cream etc. Makes about 1½ cups.

Almost-a-cake lemon pie

Vicky Musnicki *of Bowie created this dessert one time when her family wanted pie, she was short on flour and town was eight miles away. The family liked it better than the traditional pie.*

Crust:
1 cup flour
½ cup shortening
¼ cup water
⅛ teaspoon salt
½ tablespoon sugar

Filling:
1 cup sugar
2 eggs, separated
3 tablespoons soft butter or margarine
    Juice and rind of 1 lemon
3 tablespoons flour
1 cup milk

Using pastry blender, mix ingredients for crust, roll and place in 9-inch pie pan; flute edges. Prepare filling by combining sugar, beaten egg yolks, butter, lemon juice and rind, flour and milk and blend together. Fold in stiffly beaten egg whites. Pour into uncooked crust and bake at 350 degrees for 45 minutes or until done. Top should be a delicate brown. Dessert is better the second day.

Quick coffeecake

Berenice B. Foster of Tucson finds this coffeecake easy and delicious.

1¾ cups flour
¾ cup sugar
1 egg
  Milk
  Grated rind of 1 lemon
6 tablespoons melted butter or margarine
  Cinnamon and sugar

Sift flour and sugar together. Beat egg in a 1-cup measure and add milk to fill cup. Add egg mixture to flour mixture, along with the grated lemon rind and 3 tablespoons melted butter. Mix well and pour into greased oblong pan. Top batter with cinnamon and sugar. Then drizzle with remaining butter, a spoonful at a time. Bake at 350 degrees for 25 minutes or until toothpick test shows done.

Lemon-zucchini pie

Tucsonan Mary Lou Brammer first used this recipe in Iowa before she moved to Arizona.

4 cups diced peeled zucchini
¾ cup sugar
2 tablespoons tapioca
4 tablespoons lemon juice
2 tablespoons cornstarch
½ teaspoon salt
1 teaspoon nutmeg
2 tablespoons margarine
  Unbaked double-crust pastry for 9-inch pie
  Sugar

Mix all ingredients except margarine. Put filling mixture in unbaked pastry shell and dot with margarine. Put crust on top and make slits in it. Sprinkle top with sugar. Bake at 400 degrees for 10 minutes; 300 degrees for 55 minutes.

Lemon-cherry pie

Mary Nyburg of Mesa shares this family favorite that looks and tastes like a cherry cheesecake, but has no cream cheese in it.

Crust
2 cups flour
2 tablespoons sugar
2 sticks margarine
5 tablespoons water
⅓ cup slivered almonds

Filling:
1 15-ounce can sweetened condensed milk
⅓ cup lemon juice
½ teaspoon each vanilla and almond extracts
1 cup whipped cream

Topping:
2 cans cherry pie filling
1 teaspoon almond extract
4 tablespoons sugar

Combine flour and sugar and cut in margarine. Sprinkle in water and mix gently. Work in almonds lightly. Pat into 2 9-inch pie pans or 1 10-by-13-inch pan and bake for 20 minutes at 400 degrees or until slightly brown. Let cool.

Combine condensed milk and lemon juice. When thickened, add extracts and whipped cream and pour into cooled baked crusts.

Combine topping ingredients and spread on top of filling. Refrigerate pie and serve cold. Can be made several days ahead of time. Makes 2 large pies.

Lemon cream pie

With a large lemon tree in the yard,
Roberta Dean *of Phoenix needs lots of ideas for using the fruit. This pie is one of her originals.*

½ cup cornstarch
¼ teaspoon salt
2 cups sugar
3 cups boiling water
⅔ cup lemon juice
1 tablespoon lemon rind
4 egg yolks
4 tablespoons butter
1½ cups whipping cream
3 tablespoons sugar
1 10-inch pie crust, baked
Lemon slices

Mix cornstarch, salt and 2 cups sugar thoroughly. Add boiling water, lemon juice and rind.

Cook over medium heat, stirring constantly until thickened. Remove from heat, add small amount of the hot mixture to the 4 lightly beaten egg yolks. Add more of the mixture gradually until completely mixed. Return to heat. Cook for 1 minute longer. Add butter and cool.

Whip the whipping cream and beat in the 3 tablespoons sugar. Pour ½ of the lemon mixture into the 10-inch pie pan. Fold ½ of the whipped cream into the remaining lemon mixture and pour on top of mixture in pie. Top with remaining whipped cream. Decorate with thin lemon slices. Serves 10.

Lemon meringue pie

What every good cook needs is a basic lemon pie recipe that never fails, like this one from **Clara Lee Tanner** *of Tucson.*

1½ cups sugar
3 tablespoons cornstarch
2 lemons, grated rind and juice
4 eggs, separated
1¾ cups water
2 tablespoons butter
1 9-inch baked pie shell
2 tablespoons sugar

Mix 1½ cups sugar and cornstarch in top of double boiler. Add lemon rind and juice, beaten eggs yolks, water and butter. Cook and stir until thickened; cool slightly and pour into baked pie shell. Beat egg whites with 2 tablespoons sugar until stiff and spread lightly on pie. Bake at 350 degrees for 15 minutes.

Refrigerator cookies

"Learn to cook," **Lucille Young,** *Tucson, was advised some years back when she got married. One of the recipes she started with is this refrigerator cookie.*

1 cup softened margarine
½ cup white sugar
½ cup brown sugar, packed
3 tablespoons lemon juice
1 egg
4 cups sifted flour
¼ teaspoon soda
1 tablespoon shredded lemon peel
½ cup finely chopped pecans

Cream together margarine and sugars. Add lemon juice and egg and beat well. Sift flour and soda together and add to creamed mixture. Add lemon peel and pecans and mix until well blended. Shape into 4 rolls, 2 inches in diameter, and wrap in foil. Chill until firm. Cut into ⅛-inch slices and bake on cookie sheet at 400 degrees for 9 minutes or until lightly browned. Makes 7 dozen.

Lemon bars

This cookie recipe is from from Tucsonan **Christopher Aubrey,** *who enjoys baking them for friends. His interest in cooking has been helpful because of a kidney transplant that keeps him from being physically active.*

2¼ cups sifted flour
½ cup sifted confectioners' sugar
1 cup margarine or butter
4 eggs
2 cups granulated sugar
⅓ cup lemon juice
½ teaspoon baking powder
Additional powdered sugar

Mix 2 cups flour with ½ cup confectioners' sugar. Cut in butter until it resembles fine crumbs. Press into 9-by-13-by-2-inch pan and bake for at 350 degrees for 20 minutes. While crust is baking, beat eggs until foamy. Gradually beat in granulated sugar and lemon juice until smooth and fluffy. Sift ¼ cup flour with baking powder and stir into egg mixture. Pour over baked crust and return to oven for additional 25 minutes. Remove from oven. Cool in pan and sift additional confectioners' sugar over top. Cut into bars. Makes about 40.

Lemon marinade for vegetables

Alice Washburn *of Tucson sends this simple lemon mixture that is good with vegetables.*

Rounds of squash, quartered onions, cherry tomatoes, etc., as desired
1 cup oil
1 teaspoon garlic salt
⅛ teaspoon pepper
¼ cup lemon juice

Parboil briefly the squash, onions and other vegetables, as needed. Mix remaining ingredients and pour over vegetables for an hour or two. Thread on skewers and grill. Makes 1¼ cups marinade.

Lemon barbecue sauce

Larry MacMillan *of Tucson sends a barbecue sauce to use with beef, pork or chicken.*

2 tablespoons butter
1 onion, coarsely chopped
2 cloves garlic, minced
½ cup chopped celery
¾ cup water
1 cup catsup
2 tablespoons vinegar
2 tablespoons each lemon juice, Worcestershire sauce, brown sugar
1 teaspoon each dry mustard and salt
¼ teaspoon black pepper

Sauté onion and garlic in butter until lightly browned. Add other ingredients and cook for 20 minutes.

Surprise lemon- zucchini jam

This recipe came to **Ruth Weimer** *of Tucson from a California friend who got it from someone she knew in Pea Ridge, Ark.*

6 cups peeled and grated zucchini
6 cups sugar
½ cup lemon juice
1 cup crushed pineapple
1 6-ounce package apricot-flavor gelatin

Put zucchini in large pan, covered, with no added liquid. Over low heat bring zucchini to a boil and cook slowly for 15 minutes. Add sugar and boil 6 more minutes. Add lemon juice and pineapple. Boil 6 more minutes. Remove pan from heat and add the gelatin. Stir to dissolve thoroughly. Pour into sterilized glasses and seal.

Limes

Botanical name: *Citrus aurantifolia*

Delightful fruit that need protection from frost, limes are best suited for home orchards in the Salt River Valley and the Yuma Mesa. In Tucson they require protected areas. Some lime trees are grown in containers.

Limes have culinary uses far beyond drinks, as the recipes in this section suggest. Like lemons, when added to foods, limes are great for low-sodium diets.

Perhaps the most famous story about limes dates back to the days of the British sailing ships, when sailors were nicknamed "limeys" because they were issued rations of high vitamin C lime juice to prevent scurvy.

Limes often arrive at the grocery store with green skins, but they are still edible after the skins have ripened to yellow, so long as the fruit is firm. The same holds true with home-grown limes. Bearss, a Persian lime, and Mexican are the recommended varieties.

Low-calorie lime with fish fillets

Fresh limes brighten up the flavor of fish fillets. In this dish to serve three, there are only 130 calories per person!

1 pound ocean-perch fillets
2 tablespoons melted butter
1 tablespoon lime juice
1 tablespoon chopped parsley
¼ teaspoon salt
   Pinch pepper
   Paprika
   Lime wedges

If using frozen fish, thaw before cooking, Skin fillets and place on greased broil-and-serve platter. Combine remaining ingredients except paprika and lime wedges. Pour over fillets and let stand for 30 minutes. Broil about 4 inches from heat for 8 to 10 minutes or until fish flakes easily with a fork. Sprinkle with paprika. Serve with lime wedges. Serves 3. Recipe doubles easily to serve 6.

Steak salad olé with lime dressing

Shari de Long *of Phoenix finds this a great "planned over" idea: When the grill is fired up, cook extra steak for use next day in a main-dish salad.*

2 barbecued beef steaks
1 mild onion, thinly sliced
½ cup olive oil
2 tablespoons lime juice
2 tablespoons wine vinegar
  Salt and pepper
  Lettuce leaves
1 small head lettuce, shredded
2 tomatoes, cut in wedges
1 large avocado, sliced
  Parsley or coriander sprigs
2 limes, cut in wedges
  Buttered flour tortillas

Trim all fat from cold cooked steak; cut meat in thin slices, across the grain. Combine meat with onion, oil, lime juice, vinegar and salt and pepper to taste. Place lettuce leaves in wide, shallow bowl; top with shredded lettuce. Arrange meat and onions (lift from dressing with slotted spoon), with tomatoes and avocado on top. Pour remaining dressing over salad. Garnish with sprigs and serve with lime wedges, along with buttered flour tortillas. Serves 4.

Lime French dressing

The special piquancy of lime juice makes French dressing more interesting.

Combine and shake well ½ cup salad oil, ¼ cup lime juice, ¼ cup lemon juice, ½ teaspoon salt, a few grains cayenne and 2 tablespoons sugar or honey. Makes 1 cup.

Frosty lime parfait

*This attractive desert, easily made, is great for entertaining, **Roberta Dolph** of Tucson has found. Serve it in your prettiest parfait glasses or in glass dessert dishes.*

1 3-ounce package lime gelatin
1 cup hot water
1 8-ounce package cream cheese
¼ cup sugar
3 tablespoons lime juice
3 tablespoons orange juice
2 tablespoons green creme de menthe
1 cup cream, whipped
  Fresh mint leaves

Dissolve gelatin in hot water. Cool. Soften cream cheese and blend with gelatin, along with other ingredients in the order listed. Pour into individual serving bowls and chill until firm. Decorate with fresh mint leaves. Serves 8.

Lime tutti-frutti

Lime juice adds its special flavor to a cream-cheese dessert sauce that tops bananas and avocados.

1 8-ounce package cream cheese, softened
  to room temperature
½ cup milk
2 tablespoons lime juice
1 tablespoon honey
¼ teaspoon ground cinnamon
4 cups sliced bananas (about 4 bananas)
1 avocado, peeled and sliced
2 tablespoons toasted coconut

Beat together in a bowl the cream cheese, milk, lime juice, honey and cinnamon. Add bananas, mix to coat well. Cover. Chill. Just before serving, arrange avocado slices on platter, spoon banana mixture over avocado. Sprinkle with toasted coconut. Serves 4 to 6.

Lime-zucchini marmalade

Combine zucchini with lime juice and grated lime peel for this unusual marmalade.

4 cups coarsely shredded zucchini (a little over 1 pound)
2 cups water
½ cup lime juice
1 1¾-ounce package powdered fruit pectin
5 cups sugar
3 tablespoons grated lime peel

In large Dutch oven combine zucchini, water and lime juice. Bring to boil. Boil gently for 10 minutes. Stir in pectin. Return to boil. Stir in sugar and lime peel. Return to hard rolling boil that cannot be stirred down. Boil, stirring constantly, for 2 minutes.

Remove from heat. Stir for 5 minutes. Ladle into hot sterilized glasses to within ½ inch of top. Wipe edges with damp cloth. Seal with paraffin and cover or seal according to manufacturer's instructions. Store in cool place. Makes about 6 ½-pints.

Lime-cream cheese filling or fruit dip

Try this attractively seasoned cream-cheese mixture as the filling for fruits: Soften 8 ounces cream cheese and combine with 1 tablespoon each of honey and lime juice. Fold in a few gratings of lime rind and a sprinkling of nutmeg. Also good as a dip for fruits.

Key lime pie

Betty Accomazzo *of Laveen sends us this version of the popular lime pie.*

1 14-ounce can sweetened condensed milk
½ cup lime juice
3 eggs, separated
1 baked 9-inch pie crust
¼ teaspoon cream of tartar
6 tablespoons sugar

Combine milk and lime juice in medium-sized bowl. Blend in beaten yolks and turn into cooled crust. Beat egg whites with cream of tartar until soft peaks form. Beat in sugar, 1 tablespoon at a time, until whites form stiff, glossy peaks. Spoon over filling, spreading to edge of crust to seal. Bake at 325 degrees for 15 minutes or until golden. Cool on wire rack to room temperature.

Mandarins

Botanical name: *Citrus reticulata*

Though it resembles the orange in color and size, the mandarin is a distinctive form of citrus, and the tree is as ornamental as a home grower could wish. This is the fruit that furnishes the flavorful sections for commercial canning.

The Tucson area can produce mandarins, but they are usually smaller and less sweet than those grown in Phoenix and Yuma. The most popular type of mandarin is the **tangerine**, but all mandarins have the same easy-to-slip-off peel. The fruit got its "tangerine" designation because the first ones known in Europe came from Tangiers early in the 19th century.

Mandarins were cultivated in Japan more than 2,000 years ago, and long before anything was known about vitamin C, the Japanese were eating mandarins as a remedy for colds.

The mandarin crosses nicely with other fruits. The **tangor** is a cross with the sweet orange; the **tangelo**, a cross with grapefruit. If that isn't confusing enough, there are various varieties of tangelos, such as Minneola, and of tangors, such as Temple. For the home orchard, the favorite choices of mandarins include the Kinnow and the tangerine Clementine. Both are ready for eating from December to February.

Mandarin marmalade

Here's an old-fashioned marmalade recipe that can help use up a bumper crop of mandarins.

3 pounds mandarins (tangerines)
  Sugar
3 large lemons

Cut fruit in quarters but do not remove peel. Slice very thin, removing all seeds. Add finely shredded or sliced lemons. Measure fruit and add 5 times as much water. Boil until quantity is reduced nearly half — from 1 to 1¼ hours. Using 2 to 3 cups of mandarin stock at a time, add ¾ cup sugar to each cup boiling fruit and continue boiling until it gives the jelly test of thick heavy drops from the side of a spoon. This will take 10 to 20 minutes. Pour into sterilized glasses and when cool, seal with paraffin. Makes about 2 dozen small glasses.

Mandarin-chicken luncheon salad

Fresh mandarins (tangerines) give a special touch to chicken salad.

½ cup mayonnaise
  Grated peel of 1 mandarin orange
  (tangerine)
¼ teaspoon curry powder
¼ teaspoon seasoned salt
2 cups cooked chicken, cubed
2 mandarin oranges (tangerines)
  peeled, segmented and seeded
¼ cup chopped celery
¼ cup chopped green pepper
¼ cup toasted slivered almonds
  Salad greens

In large bowl, combine mayonnaise, mandarin peel, curry powder and seasoned salt. Stir in remaining ingredients except salad greens. Chill. Serve on salad greens arranged on 4 plates.

Tangerine coleslaw

Tangerine juice, peel and segments make this unusual cabbage salad a party treat.

¼ cup salad oil
  Grated peel of ½ a tangerine
¼ cup tangerine juice
  Juice of ½ lemon
2 tablespoons honey
1 small head cabbage, shredded
  Segments of 3 tangerines, cut
  in half and seeded
½ cup raisins
¼ cup chopped nuts

Combine oil, tangerine peel and juice, lemon juice and honey in jar and shake well. In large bowl, combine cabbage, tangerine pieces, raisins and dressing. Chill. To serve, toss with chopped nuts. Serves 6.

Oranges (sweet)

Botanical name: *Citrus sinensis*

The orange got its name, say some food historians, from the Sanskrit word for the fruit, *naranga*. Others suggest it is derived from the Old French word for gold, *auranja*. The fruit is believed to be the golden apple of classical mythology.

Oranges come sweet or sour, and unquestionably, the sweet types are our most beloved citrus. The sweet ones include Valencia and similar juice types, Washington navels and similar seedless types, and blood oranges, not very well known in this country.

Most everyone calls juice oranges "valencias," and oranges of this type generally prove more satisfactory for home growers. In the Sonoran Desert, navels tend to split and drop from the tree. say some horticulturists. Their advice: Stick with the juice oranges.

Though the name does the fruit no credit, the sweet orange known as the "blood orange" is the delight of gourmets. The origin of the name is obvious: The pulp of this citrus is definitely on the red side.

Blood oranges are not grown commercially in the United States, partly because the color of the juice is variable when canned or frozen. The fruit is popular in Mediterranean countries, and travelers to that part of the world find the reddish juice the favorite breakfast beverage at restaurants and hotels.

Blood oranges have the flavor of oranges with a hint of raspberries or grapes. There may be a red blush on the peel. One taste of their exotic flavor and their unfortunate name is forgotten.

Better choices for home orchards among juice oranges are the Valencia and the Arizona Sweets: Marrs, Pineapple, Diller, Hamlin and Trovita; among the navels, Washington Navel and Robertson Navel; among the blood oranges, Sanguinelli and Moro.

Glazed pork chops with sweet potatoes

Using fresh citrus in cooking is popular in Yuma. **Vickie Steinfelt** *finds this pork and sweet potato dish a flavorful dinnertime choice.*

4 good-sized pork chops
1 tablespoon shortening
1 clove garlic
3 tablespoons sugar
2 tablespoons flour
1 teaspoon salt
½ cup water
½ cup orange juice
2 tablespoons lemon juice
½ teaspoon rosemary
2 1-pound 7-ounce cans sweet potatoes

Melt shortening in skillet. Add garlic and chops. Brown chops on both sides. Remove chops and discard garlic. Combine sugar, flour and salt and blend into drippings. Stir in water, orange juice, lemon juice and rosemary. Stir and cook until thickened. Place chops in casserole and pour sauce over them. Cover and bake at 350 degrees for 45 minutes. Drain sweet potatoes and add to casserole, spooning sauce over them. Cover and bake for 15 more minutes or until chops are tender. Serves 4.

Sunshine beef bake

This main dish is a version of Phoenician **Rosemary Bradley's** *first-place entry in the 1977 Arizona Beef Cookoff.*

2 pounds beef round steak, cut ½-inch thick
¼ cup flour
½ teaspoon each salt and garlic salt
2 cups soft bread cubes
¼ cup chopped green pepper
1 tablespoon instant toasted onion
⅓ cup beef bouillon
1 teaspoon Worcestershire sauce
2 oranges
   Sunshine orange sauce (see below)
   Orange slices, parsley for garnish

Cut round steak into 6 serving-size pieces. Dredge meat in combined flour, salt and garlic salt. Combine bread cubes, green pepper and onion and stir in bouillon and Worcestershire sauce. Pare oranges (cutting round and round) in long spirals. Cut orange segments in small pieces and stir into stuffing mixture.

Divide stuffing into 6 equal portions and place 1 portion on top of each steak. Fold opposite edges of each steak up and overlap on top. Secure with small round wooden picks and place, seam side down, in greased shallow, 2-quart casserole. Bake, uncovered, at 350 degrees for 30 minutes. Remove from oven and place spirals of orange peel around meat.

Prepare sunshine sauce and spoon over beef rolls. Cover tightly, return to oven and bake for 2 hours until meat is tender. Serve on warm platter, garnished with orange slices and parsley, if desired.

Sunshine sauce: Shake well 1 cup orange juice, ½ cup slivered almonds, ¼ cup cooking oil, 3 tablespoons soy sauce and 2 tablespoons French dressing.

Orange-pineapple chicken breasts

Orange juice and pineapple chunks make this main dish especially attractive.

3 whole broiler chicken breasts,
   boned, skinned, and cut in 2-inch pieces
¾ teaspoon salt
¼ teaspoon ground ginger
2 tablespoons vegetable oil
1 small garlic clove, minced
1 8¼-ounce can pineapple chunks,
   undrained
1 cup orange juice, divided
1 envelope instant chicken bouillon
2 tablespoons wine vinegar
½ cup sliced celery
1 small green pepper, cut into ¼-inch strips
1 small onion, sliced
1 small tomato, cut in wedges
2 tablespoons soy sauce
1 tablespoon sugar
3 tablespoons flour

Sprinkle chicken with salt and ginger. Heat oil in large skillet over medium heat. Add chicken and garlic. Cook 5 minutes. Add liquid from canned pineapple, ¾ cup orange juice, bouillon and vinegar. Cover. Simmer 10 minutes. Add celery, green pepper and onion. Cover. Cook 5 minutes longer. Add tomato wedges and pineapple chunks.

In small bowl, blend together soy sauce, sugar, flour and remaining ¼ cup orange juice. Add to skillet. Cook, stirring constantly, until mixture thickens and boils. Cook 1 minute longer.

Serve over hot, cooked rice, if desired. Serves 4.

Beefed-up breakfast muffins

For brunch, **Karen Christensen** *of Phoenix suggests baking these muffins in advance and storing in the refrigerator or freezer. To serve, heat in toaster oven or microwave.*

1 pound ground beef
1 teaspoon salt
½ teaspoon soda
¼ teaspoon ground cumin
2 cups biscuit mix
1 egg, beaten
⅔ cup orange juice
1 teaspoon grated orange peel

Lightly brown ground beef in frying pan. Remove meat to absorbent paper; sprinkle salt over meat. Stir soda and cumin into biscuit mix. Combine egg, orange juice and orange peel; add to biscuit mix and stir until just combined. Fold meat into batter. Place paper liners in 12 medium muffin cups and add ¼ cup batter to each. Bake at 400 degrees for 25 to 30 minutes. Makes 12.

Orange chicken

A sweet-sour orange sauce makes this chicken special.

1 3-pound broiler-fryer, cut up
½ cup unbleached flour
¼ teaspoon garlic powder
3 tablespoons oil
¾ cup orange juice
2 medium onions, thinly sliced
⅓ cup soy sauce
⅓ cup apple cider vinegar
2 tablespoons honey
2 tablespoons water
1 large green pepper, seeded and sliced
1 cup sliced mushrooms

Combine flour and garlic powder and dredge chicken. In frying pan, brown chicken slowly in oil. Add orange juice and onions. Cover and simmer for 20 minutes.

Mix together soy sauce, cider vinegar, honey and water and pour over chicken. Add pepper and mushrooms and cover again. Bake at 325 degrees for about 25 minutes, or until chicken is tender. Serve in shallow bowls. Sauce may be thickened, if desired. Serves 5 or 6.

Orange

Citrus crumb coating

To make a flavorful coating for breading pork chops or chicken, combine 1 cup fine dry crumb, 1 teaspoon grated orange peel and ½ teaspoon basil and ½ teaspoon salt. Dip chops or chicken pieces first in beaten egg, if desired, then in crumb coating and brown in skillet. Add small amount of water, cover and simmer until tender, about 45 minutes.

Orange duckling

For speedier cooking, this duckling with orange sauce is baked in quarters.

1 4½- to 5-pound duckling
½ teaspoon salt
3 small bay leaves
½ cup each orange juice and water
1 tablespoon each sugar and
   cornstarch
1 chicken bouillon cube, crushed
2 tablespoons orange liqueur
1 tablespoon shredded orange rind
   Cooked peas for 4
   Sliced oranges, halved

Defrost duckling, cut in quarters, wash, drain and dry. Combine salt and 2 bay leaves, finely crushed, and rub over duckling. Arrange quarters, skin side up, on rack in roasting pan. Bake at 350 degrees until meat on drumstick is tender, about 2 hours. Turn quarters often, ending with skin side up.

To prepare sauce, combine orange juice and water, sugar, cornstarch and bouillon cube; stir until free of lumps. Add remaining bay leaf. Cook until thickened, stirring constantly. Simmer, uncovered, about 5 minutes. Add orange liqueur and rind; heat thoroughly. Arrange duckling and peas on serving platter with orange slices for garnish. Serve topped with sauce. Makes 4 servings.

Orange pancake or dessert sauce

Try this sauce on waffles, pancakes or cake.

1 cup sugar
4 teaspoons cornstarch
1½ cups cold water
4 teaspoons margarine
   Grated rind of 1 large orange
   Juice of 1 large orange

Combine sugar, cornstarch and water in top of double boiler; heat over boiling water, stirring, until sauce thickens. Remove from heat and add margarine, rind and juice. Let cool to room temperature and stir thoroughly. May be chilled before serving. Makes about 1½ cups.

Orange pancakes

These pancakes are great for Sunday breakfast with honey.

2 large eggs
3 tablespoons oil
2 cups flour
½ teaspoon soda
½ teaspoon salt
   Grated rind of 1 orange
   About 2 cups orange juice
1 cup yogurt

Beat eggs with oil. Sift flour, soda and salt together; add to eggs along with orange rind. Gradually beat in orange juice, adding more juice if batter is thicker than you like. Beat until smooth with wire whisk.

Heat griddle or skillet; fry pancakes until golden brown. To serve, place a pancake on each of 4 warm plates; top with ¼ cup yogurt, then with another pancake. Serve with orange sauce. Makes 4 servings, about 8 pancakes.

Cranberry-orange candle salad

Phoenician **Ruth C. Brunton** *shares this festive holiday idea, to be served with lighted candles.*

1 1-pound can whole-cranberry sauce
1 3-ounce package orange gelatin
1 cup boiling water
¼ teaspoon salt
1 tablespoon lemon juice
½ cup mayonnaise
1 orange, peeled and diced
¼ cup chopped nuts

Heat cranberry sauce, strain and set berries aside. Dissolve gelatin in hot juice and boiling water. Add salt and lemon juice and chill until thickened enough to mound slightly when dropped from a spoon. Beat in mayonnaise until light and fluffy. Fold in cranberries, orange pieces and nuts. Divide mixture evenly into 8 6-ounce fruit-juice cans. Chill 4 hours or longer. Unmold on beds of lettuce, standing on end like "candles." Cut thin birthday candles in half to shorten and insert in the gelatin "candles." Just before serving, light candles. Serves 8.

Mediterranean orange salad

An orange salad served with fresh mint leaves is a pretty dish from **Angela C. Parker** *of Phoenix.*

5 large navel oranges
3 tablespoons salad oil
1 tablespoon wine or cider vinegar
¼ teaspoon salt
⅛ teaspoon black pepper
¼ clove garlic, crushed
¼ teaspoon oregano
Fresh mint leaves

Peel and section oranges. Place in bowl and chill. Beat together the oil, vinegar, salt, pepper and garlic and chill. Just before serving, sprinkle oranges with oregano and a little black pepper. Strain garlic from dressing and pour over the oranges. Garnish edge of bowl with fresh mint.

Orange bread

Tucsonan **Esther Hurcombe's** *bread has a lot going for it: It makes a great gift, freezes well, is very good toasted.*

3 cups sifted flour
4 teaspoons baking powder
¾ teaspoon salt (optional)
⅓ cup soft shortening
¾ cup milk
1 medium orange, diced (do not peel)
1 egg
1 cup sugar

Sift flour, baking powder and salt into a mixing bowl. Place shortening, egg and milk in electric blender container. Add orange (seeds removed) and sugar. Blend until orange is finely chopped (about 1 minute). Pour over flour mixture and stir until flour is moistened. Pour into greased large loaf or 2 small loaf pans. Bake at 350 degrees for about 1 hour. Cool on its side on wire rack. Let stand overnight before slicing.

Options: Sprinkle with cinnamon and sugar before baking; add ½ cup finely chopped nuts to batter.

Orange pecan bread

Use a food processor on this one, if you have it; otherwise a good food chopper, recommends Tucsonan **Virginia B. Selby.**

1 to 2 oranges
½ cup butter
⅓ cup sugar
1½ teaspoons salt
1 package yeast
1 egg
½ cup chopped pecans
2½ cups flour (about)

With vegetable peeler, peel oranges. Remove as much of the white portion as possible and place orange peeling and pulp in food processor or chopper. There should be 1 cup of finely chopped pulp, peeling and juice. Combine with sugar, butter and salt and bring to a boil. Cool to lukewarm, add yeast and beat to blend well. Add beaten egg, pecans and enough flour to make a stiff dough.

Knead on well-floured board for 5 minutes, adding flour as needed to keep from sticking. Cover with cloth and let rise in warm place until doubled in bulk (about an hour). Knead again for 1 minute, shape into loaf and put into greased loaf pan, 10-by-5-by-3½ inches. Cover and let rise until a little more than double. Bake at 375 degrees for about 40 minutes. Cool on rack.

Orange-date bread

Serve Tucsonan **Bette Henry's** *orange-date bread plain, sprinkled with powdered sugar, or spread with cream cheese.*

1 medium orange
    Boiling water
1 cup ground dates
2 tablespoons shortening
1 teaspoon vanilla
1 slightly beaten egg
2 cups flour
1 teaspoon baking powder
¼ teaspoon salt
½ teaspoon baking soda
1 cup granulated sugar
½ cup chopped pecans

Place juice of orange in measuring cup and fill with boiling water to 1 cup. Grind orange rind and dates. Place ground dates and rind, shortening and hot diluted juice in medium-size mixing bowl. Add vanilla and beaten egg. Combine dry ingredients and add to fruit mixture. Mix well, then add nuts. Put in greased and floured loaf pan. Bake at 350 degrees for 1 hour or until done when tested with a toothpick. Cool in pan. Bread freezes well. Slice and serve plain or top with cream cheese.

Orange stuffing

Ideal for capon or a small turkey, this bread stuffing is seasoned with orange juice and peel.

½ cup chopped onion
½ cup chopped celery
¼ cup butter
½ teaspoon salt
¼ teaspoon poultry seasoning
    Dash pepper
1 tablespoon grated orange peel
¼ cup orange juice
4 cups dry bread cubes (half white, half whole wheat)

Cook onion and celery in butter until tender. Sprinkle seasonings over bread cubes. Combine onion mixture, orange peel and juice. Pour over bread cubes. Toss lightly.

Orange ring

No frosting is necessary in this cake recipe from **Eleanor Beaulac Blattner** *of Tucson.*

2 cups flour
1 teaspoon baking powder
1 teaspoon soda
1 teaspoon salt
1 cup sugar
¾ cup shortening
¾ cup dates or raisins
    Juice and rind of 1 orange
2 eggs
¾ cup sour milk
1 teaspoon vanilla
½ cup chopped nuts

Mix and sift dry ingredients; cut in shortening. Put raisins and orange rind through food chopper. Mix well with flour and shortening; add eggs, then milk and vanilla, and stir until smooth. Stir in nuts. Bake in greased tube pan at 350 degrees for 1 hour. Remove from oven and immediately pour orange juice over the top. Let stand a few minutes for orange juice to penetrate through the cake and for the top to dry.

Orange date cake

Taken from a Graham (Texas) Methodist cookbook of the '50s, this cake is great for the holidays, Tucsonan **M. Louise Witkowski** *has found.*

1 cup butter
2 cups sugar
1 teaspoon soda
½ cup buttermilk
4 eggs
3½ cups flour
2 tablespoons grated orange peel
1 cup chopped dates
1 cup chopped pecans

Cream butter and sugar. Add soda mixed with buttermilk. Add eggs and flour alternately. Add grated orange peel, chopped dates and nuts. Bake in large angel cake pan, well greased and the bottom covered with waxed paper, also greased. Bake at 325 degrees for 1¼ hours or more, testing with a toothpick to see whether it is done. While still hot, baste with orange sauce, continuing to ladle sauce over cake until all is absorbed.

Sauce: Combine 2 cups powdered sugar, 1 cup orange juice and 2 tablespoons grated orange peel.

Orange-raisin cake

Tucsonan **Helen Larsen** *was given this cake recipe by her sister, whom she describes as "the best cook in Southern Utah."*

1 large orange, peeled
1 cup raisins
1 cup pecans or walnuts
2 cups flour
1⅓ cups sugar
1 teaspoon salt
1 teaspoon soda
½ cup shortening
1 cup milk
2 eggs
    Orange-powdered sugar frosting

Grind orange, raisins and nuts. Combine flour, sugar, salt and soda in a sifter. Sift into a large bowl. Add milk, shortening and eggs, and beat in electric mixer for 2 minutes at low speed. Stir in orange mixture. Bake as a sheet cake at 350 degrees in greased 13-by-9-inch pan for 40 to 50 minutes, in larger pans for 30 minutes. Frost when cool.

Frosting: 1 regular-size box confectioners' sugar, ⅓ cup butter, 2 or more tablespoons half and half, 1 teaspoon vanilla and 2 teaspoons grated orange rind. Blend well.

Orange sauce

Robin DeBell of Tucson recommends this low-calorie citrus sauce for serving cold as a dessert sauce or warm over broiled chicken.

    3 oranges
    3 teaspoons finely grated orange peel
    ½ cup water
    ½ teaspoon vanilla
    ⅛ teaspoon ground cloves
    2 teaspoons fructose sweetener

Grate peeling (orange-portion only); then peel and dice orange pulp over a bowl to catch the juice. In small saucepan, combine juice, grated peel, water, vanilla and cloves. Bring to boil and boil 2 minutes. Add orange pieces and simmer 8 to 10 minutes. Cool and refrigerate. When cold, mix in fructose. Good over fruit or dessert crepes or warm and add to broiled chicken. Serves 4, 40 calories each.

Orange custard

An old-fashioned egg custard, spiced with orange juice, is a delightful dessert.

    2 tablespoons butter or margarine
    ½ cup honey
    ¼ cup unbleached flour
    6 eggs, separated
    2 cups orange juice
      Dash of salt

In a mixing bowl, cream together butter or margarine and honey. Stir in flour until smooth and beat in egg yolks.

Add orange juice, a dash of salt and mix well. Beat egg whites until stiff and fold into orange juice mixture. Pour into a buttered 1½-quart baking dish and set in a shallow pan of hot water. Place in a 350-degree oven for 1 hour, or until knife inserted into the middle of the custard comes out clean. Cool slightly and spoon into dessert dishes. Serve with whipped cream, if desired. Serves 6.

Oranges

Orange soufflé

Ann White of Ajo usually makes this dessert with oranges, although other fresh fruit, pureed, can be used.

    1 envelope unflavored gelatin
    ¼ cup water
    3 eggs, separated
    1 cup sugar
      Grated rind and juice of 2 oranges
    ½ cup heavy cream, slightly whipped
      Fruit, nuts or whipped cream

Soften gelatin in water. Mix yolks and sugar in top of double boiler and cook over boiling water, beating until light in color and thickened. Remove from heat, stir in softened gelatin and stir until dissolved. Beat in the juice and rind and continue beating until cool. Beat egg whites until stiff; whip cream lightly. When fruit mixture is cold, place pan over ice cubes and stir until mixture begins to set. Fold in cream and egg whites and pour into mold. Serve garnished with nuts, fruit or whipped cream. Serves 4 to 6.

Tangy orange sherbet

This sherbet from Tucsonan **Mae Criley** *is made without gelatin. She likes to prepare it a day ahead when entertaining at dinner.*

Juice of 2 oranges
Juice of 2 lemons
Additional juice (pineapple or other)
1¾ cups sugar
2 cups whipping cream
1 tablespoon grated orange rind
1½ teaspoon grated lemon rind

To the orange and lemon juices add enough pineapple juice to make 2 cups. Combine sugar and cream. Blend in juices and rind. Pour into tray and partially freeze. Remove from freezing compartment and beat vigorously (food processor can be used for this). Return to freezing compartment and freeze until firm. Serves 12.

Orange pudding

Mary Loso *of Tucson enjoys making this fresh orange pudding, one of her special recipes.*

3 oranges, seeded
½ cup sugar
2 eggs, separated
1 tablespoon cornstarch
2 cups milk
1 tablespoon sugar

Peel and cut oranges in small pieces. Put in shallow, small pan or ovenproof casserole. Sprinkle with ½ cup sugar. Beat yolks well; add cornstarch. Let milk come to a boil, add the egg yolks and cornstarch and cook and stir until thick. Let custard cool. Pour over oranges. Beat whites until stiff, and add 1 tablespoon sugar. Spread over oranges. Bake at 350 degrees until golden brown. Serves 4.

Honey orange cooler

When you have fruit juices on hand, whip up this flavorful cooler, suggests **Barbara Stockwell** *of Arivaca. A touch of honey gives the right added sweetening. It's a mixture a mother can really trust.*

Mix well and chill: 1 cup orange juice, ¼ cup lemon juice, 2 cups water, 2 cups grape juice and 4 tablespoons honey.

Carrot-orange marmalade

Rose Gillin *of Tucson makes a flavorful marmalade with oranges and carrots.*

Scrape 6 carrots, dice and cook until tender in as little water as possible. Slice 3 oranges very thin and add juice and grated rind of 1 lemon. Combine carrots and fruit, measure into saucepot and add ⅔ as much sugar as the total of carrots and fruit. Heat, stirring until sugar is dissolved. Cook rapidly until thick and clear. Pour into hot clean jars and seal. Makes 6 6-ounce jars.

Orange marmalade

This recipe, which **Hazel Owens** *of Tucson brought with her many years ago from New Orleans, makes beautiful marmalade with Arizona fruit.*

8 oranges, on the green side
2 grapefruit
3 lemons
Sugar

Remove seeds and chop fruit or grind coarsely. To each pound of fruit, add 1 quart of water. Let stand overnight, then boil until tender. Cover and let stand overnight again. Weigh again. To each pound of fruit add 1½ pounds sugar. Boil and stir until it jells, about 1 hour. Pour into sterilized jelly glasses and seal.

Oranges (sour)

Botanical name: *Citrus aurantium*

Skimpy on juice and long on ornamental qualities, the sour orange is all too often overlooked in the kitchen.

The fruit begins to ripen in late fall, and stays on the tree for many months. Finally, after the bright globes have fulfilled their decorative role, thousands fall on the ground, where they remain until carted off or children pick them up to play catch with.

It's a sad sight, because the trees furnish not only beauty but aromatic, sour juice that can be used with great success in drinks, marmalade and desserts, as well as other ways. They may also be used while still small and green.

Those who admire sour oranges suggest treating the juice as if it were lemon juice, and adding sweetening as needed. The sour orange has a rind much thicker than the sweet orange, and may yield less than 3 tablespoons of juice. There are numerous seeds.

The most popular sour orange is the Seville. Chinotto and Bouquet are other successful types.

Yucatan barbecued chicken

This recipe from **Leonie Hulse** *of Tucson is one she enjoyed on a trip to the Mexican state years ago. It can be made with or without the seasoning annatto, a reddish brown seed of the annatto tree.*

½ teaspoon salt
¾ cup juice of Seville sour oranges
12 peppercorns
4 cloves garlic
1 tablespoon annatto, optional
½ teaspoon oregano
¼ teaspoon cumin
½ teaspoon allspice
3 Cornish hens, cut in half
   Banana leaves, optional
6 thin slices onion
6 slices tomato
1 small can jalapeño chiles

Place salt, orange juice, peppercorns, garlic, herbs and spices in electric blender and blend until smooth. Place the cut Cornish hens in a shallow dish and cover with marinade, rubbing well into each piece. Refrigerate for 24 hours, turning once or twice.

Wrap individual servings in foil packets as follows: On foil, place banana leaf, then ½ Cornish hen, top with slice of onion, slice of tomato and slice of jalapeño. Distribute remaining marinade between the 6 packets. Seal well. Bake at 325 degrees for about 2½ hours. Serves 6.

Seville pork steaks

When there are sour oranges on the tree, you can be ready with marinade for pork chops in a hurry, says Tucsonan **Tres English.**

2 or 3 pork steaks or chops
½ cup sour-orange juice, seeds removed
1½ tablespoons Worcestershire sauce
1½ tablespoons brown sugar
½ teaspoon salt

Brown pork steaks or chops in skillet. Combine sour-orange juice with Worcestershire sauce, brown sugar and salt. Pour over meat. Cover and simmer until meat is tender, about 45 minutes, turning several times and adding small amount of water as needed. Sauce is also good with chicken.

Sour orange blossoms

Sour orange bread

Gloria Thomasson *of Tucson became interested in cooking with sour oranges and tried several ideas with the fruit. Here is her flavorful bread.*

1 cup milk, scalded
⅔ cup Grape Nuts cereal
1 teaspoon grated sour orange peel
⅔ cup sugar
⅓ cup sour-orange juice, strained
½ teaspoon baking soda
⅓ cup salad oil
1 egg
2 cups all-purpose flour
½ teaspoon salt
2 teaspoons baking powder

Add cereal and peel to scalded milk; stir in sugar and set aside to cool. Add sour orange juice, baking soda, salad oil and egg. Stir to blend. Stir in flour, salt and baking powder and mix well. Pour into 8½-by-4½-inch loaf pan. Bake at 350 degrees for 1 hour; cool and wrap in foil. Store several hours or overnight before slicing.

Ornamental orange pie

Here's a method for making a sour or ornamental orange pie, using cornstarch in the filling.

1 baked pie shell
Juice of 3 or 4 sour oranges (½ cup)
2 eggs, separated
1 cup sugar
Dash salt
2 tablespoons cornstarch
¼ teaspoon cream of tartar
¼ cup sugar

To the strained juice add enough hot water to make 2 cups liquid. Combine egg yolks, 1 cup sugar, salt and cornstarch mixed with a small amount of cold water. Then add the hot water-sour orange mixture. Bring to a boil. Let cool, pour into baked crust and top with meringue.

To make meringue, beat whites with cream of tartar until frothy. Add ¼ cup sugar slowly, beating continuously until egg whites reach the stiff-peak stage. Spread on filling, sealing edges, and bake at 400 degrees until light brown (about 10 minutes).

Sour orange chiffon pie

An ardent believer in using sour oranges, **Jeanette Bideaux,** *Tucson, has often made this delightful pie with oranges from her yard.*

4 eggs, separated
1 cup sugar
½ cup juice from ripe sour oranges
1 tablespoon lemon juice (optional)
½ teaspoon salt
1 envelope gelatin
¼ cup cold water
1 tablespoon grated sweet-orange rind
1 baked pie shell

Add ½ cup sugar, sour-orange juice and salt to beaten egg yolks and cook over boiling water until of custard consistency. (Combine 1 tablespoon lemon juice with enough sour-orange juice to make the required ½ cup, if you want more zing to your pie filling.) Pour cold water in bowl and sprinkle gelatin on top of water. Add to hot custard and stir until dissolved. Add grated sweet-orange rind. Cool.

When mixture begins to thicken, fold in stiffly beaten egg whites to which the remaining ½ cup sugar has been slowly added. Fill baked pie shell and chill. Add a thin layer of whipped cream before serving. Makes 1 pie.

Early sour orange pie

Two University of Arizona experimental-foods students, **Beverly Semple and Paula Sendar,** *found that immature sour oranges, green-skinned and half grown, have a pleasing citrus flavor.*

1 unbaked pastry shell
1 tablespoon unflavored gelatin
1 cup sugar
¼ teaspoon salt
4 eggs, separated
½ cup immature sour-orange juice
¼ cup water
1 teaspoon grated peel
   Green food coloring
1 cup heavy cream, whipped

Mix gelatin, ½ cup sugar and salt in saucepan. Beat together the egg yolks sour orange juice and water. Stir into gelatin mixture. Cook and stir over medium heat just until mixture comes to a boil. Stir in peel. Add food coloring until pale green.

Chill until mixture mounds slightly when dropped from a spoon. Beat egg whites until soft peaks form. Gradually add ½ cup sugar, beating to stiff peaks. Fold gelatin mixture into egg whites. Fold in whipped cream. Spoon into cooled baked pastry shell and chill until firm. Decorate with additional whipped cream, if desired. Makes 1 pie.

Sour orange sponge pudding

Doris Schultz *of Green Valley revised a favorite recipe to come up with this idea.*

¼ cup butter or margarine
1 cup sugar
¼ cup flour
3 eggs, separated
1⅓ cups milk
   Grated rind of 1 sour orange
   Juice of 2 sour oranges
⅛ teaspoon salt

Cream butter and sugar. Blend in flour. Beat egg whites stiff and set aside. Beat yolks until thick; add milk, rind and juice and salt. Add to flour mixture. Mix well and fold in egg whites. Pour into buttered casserole. Place casserole dish in pan of hot water and bake at 350 degrees for 45 minutes to 1 hour. Serve warm or cool.

Ornamental orange cake-topped pudding

Audrey Jones *of San Manuel developed this pudding idea using the juice and rind of the ornamental or sour orange.*

2 tablespoons softened butter or margarine
1½ cups sugar
⅓ cup flour
¼ teaspoon salt
½ cup sour orange juice
1 teaspoon grated orange peel
3 eggs, separated
1¼ cups milk

Cream butter and sugar. Combine and stir in flour and salt, then the juice and peel. Mix beaten egg yolks in milk and add to orange mixture. Beat egg whites until stiff peaks form and fold into mixture. Pour into 6 to 8 custard cups; set cups in pan and fill pan 1-inch deep with water. Bake at 375 degrees for about 40 minutes. Cool slightly and turn upside down on individual serving dishes. Serve warm or chilled with whipped or sour cream.

Sour orange butter cookies

Beverly Semple and Paula Sendar *made these butter cookies in their UA experimental-foods class with immature sour-orange juice and grated rind.*

1 cup butter
1 cup light brown sugar
2 eggs, well beaten
1 teaspoon grated immature sour
  orange rind
  Juice of ½ immature sour orange
1 teaspoon cinnamon
¼ teaspoon powdered cloves
¼ teaspoon salt
2 cups flour

Cream the butter and brown sugar thoroughly. Add eggs, sour-orange rind and juice, cinnamon, cloves and salt. Mix well. Blend in flour, adding more if necessary to make dough thick enough to roll out. Chill. Roll to ⅛-inch thick and cut in desired shapes. Bake at 350 degrees until delicately brown, about 10 minutes.

Sour orange-baked apples

The special flavor of sour oranges give a special touch to these baked apples from **Audrey Jones** *of San Manuel.*

6 large tart baking apples
½ cup sugar
1 sour orange
¼ cup raisins
2 tablespoons chopped pecans
¾ cup maple syrup
1 teaspoon grated orange rind

Wash and core apples; pare apples halfway down from top and stand in shallow 9-by-13-inch baking dish. Pare skin and thick white membrane from orange and dice fruit. Toss sugar with orange pieces, raisins and pecans and spoon into centers of apples. Pour syrup over apples and sprinkle with grated peel. Bake at 375 degrees for 1 hour or until apples are tender. Baste with syrup frequently during and after baking. Serve warm or chilled.

Sour orange spiced marinade

Cloves and cardamom add the right spicy touch to this marinade for pork or chicken.

⅓ cup sour orange juice
   Water
2 tablespoons packed brown sugar
⅛ teaspoon ground cloves
1/16 teaspoon ground cardamom
2 to 3 pounds country-style ribs or
   1 broiler-fryer, cut up
   Salt

Add water to juice to make ½ cup, and stir in sugar and spices. Brown ribs or chicken pieces in small amount of fat (pork, chicken or other fat or oil), sprinkle with salt. Pour marinade over ribs, cover and simmer until tender, about 1½ hours for ribs, less for chicken, turning occasionally. Remove meat or chicken from skillet and keep warm while cooking down sauce to thicken, as desired. Serves 4 or 5.

• **See also rosy grapefruit-sour orange marmalade, Page 44.**

Sour orange-yogurt dressing

1 cup plain yogurt
2 tablespoons honey
1 teaspoon chopped chives
½ teaspoon garlic salt
   Juice of 1 sour orange

Stir to mix and serve as dressing for Western salad.

Sour orange-cranapple jelly

Geraldine Lynde *of Sun City worked out this interesting combination with juice from the ornamental sour orange trees in her yard.*

2 cups sour orange juice
2 cups cranapple juice
5½ cups sugar
1 pouch or ½ bottle liquid pectin

Mix sugar and juices thoroughly in large kettle and bring to full rolling boil. Boil for 8 minutes without stirring. Add pectin and bring to full, rolling boil. Boil for 1 minute, stirring constantly. Remove from heat, stir and skim for 1 minute. Transfer to hot clean glasses. (If prepared when humidity is under 25 percent, reduce 8-minute boiling time to about 7 minutes.) Makes 7 6-ounce jelly glasses.

Other citrus

Citron

Botanical name: *Citrus medica*

Desert nurserymen say they rarely get calls for citron trees, but some are grown in the low desert, mostly as a novelty. Citron trees are not as attractive as other citrus and are very sensitive to cold. The favorite for home growing is the Etrog.

One variety is shaped with fingerlike sections. It is known as "Buddha's hand." Another type, which sometimes has a somewhat bumpy skin and resembles a rather large lemon, was traditionally required by the Mosaic law of the Hebrews to be passed among the congregation in the Feast of Tabernacles.

There is little pulp and juice inside a citron, but the rind is very thick and aromatic. The rind is candied using the same method as for grapefruit rind.

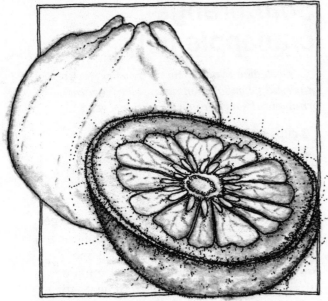

Pummelos

Cocktail citrus

This one is not a variety but a fun tree created by grafting several citrus — orange, grapefruit, lemon etc. — to a single root stock. Each area of the tree then produces its own kind of blossoms and fruit.

There is one caution: Watch the pruning, or eventually there may be only one variety of fruit on the tree.

But for those who enjoy a conversation-type tree, the cocktail citrus is just right. As one nurseryman puts it, "If you have room for just one tree and can't decide which one you want, the cocktail citrus tree may be just the thing."

Pummelos

Botanical name: *Citrus grandis*

This ancient relative of the grapefruit should not be confused with *pomelo,* which means grapefruit in Spanish.

Pummelos require the warm climate of Yuma or Phoenix, but occasionally people succeed with them in Tucson's warm microclimates. Hybrids are being grown experimentally in California.

Pummelos are the largest of the citrus fruits; there are reports of some that weigh as much as 15 pounds. They are less bitter than grapefruit, with numerous seeds. Some have a pearlike shape. They are of special interest to Chinese-Americans at the midwinter festival time. They are not grown much outside of Southeast Asia.

The fruit is also sometimes called a **shaddock,** after the English ship's captain who introduced it to the West Indies in the 17th century.

To eat a pummelo, it is necessary to peel off the rind and then remove each section of fruit from its membrane covering with a sharp knife. The flesh is light yellow to pink. For home growing, the best choices are Kao Phuang and Reinking.

Other fruits

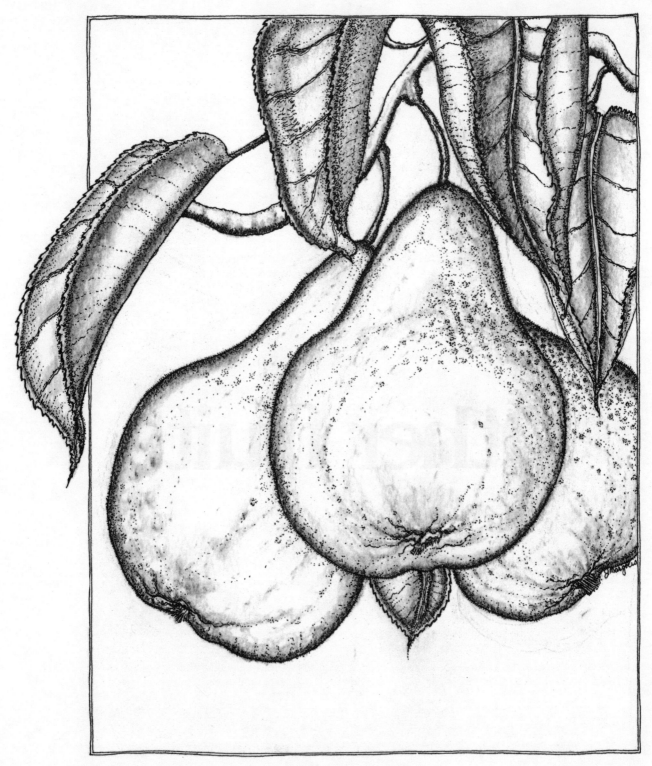

Bartlett pears do well in home orchards in some areas.

Apples

Botanical name: *Malus sylvestris*

Grow apples in the desert? Some heat resistant apples, such as the Anna developed by the Israelis, have been successful. But in the southern half of Arizona, it is in the higher altitudes of Graham, Cochise and Greenlee counties, especially, that apples do best.

One enterprising Tucson-area home grower of apples, however, reports that he has coaxed traditional trees to bear by putting ice around the roots for several weeks each winter.

Apple trees have appeal beyond the fruit: the visual pleasure of their delicate white or pink blossoms, which resemble wild roses. (The apple belongs to the rose family.)

At 3,500 to 7,000 feet, successful early varieties for home growing are Lodi, Gravenstein and Milton McIntosh; fall and winter varieties include McIntosh, Golden and Red Delicious, Jonathan, Winesap and Rome Beauty. Below 3,500 feet, besides Israeli apples, Early Summer Red and Tropical Beauty are satisfactory. Santa Cruz County can succeed with such varieties as Granny Smith and Winter Banana. Dwarf sizes are especially appropriate to home orchards. Some varieties require a companion tree as a pollinator.

Most beloved of all fruits, apples were brought to the New World by European colonists. They were carried westward by early settlers, with a special boost from Johnny Appleseed. They now grow in virtually every state.

Apples have long been considered healthful, though modern nutritional analysis shows they offer only fair amounts of some vitamins and minerals. They contain pectin and desirable fiber, and when eaten raw, help keep teeth and gums in good condition — thus keeping the dentist away, if not the doctor.

Crab apples, smaller and more acid than the standard apple, make a fast-growing lawn tree. They are enjoyed more often for their springtime blossoms than for their fruit, although they can be eaten raw or used in jelly and pickles.

Depending on variety, crab apple trees grow to heights of 15 to 40 feet. For the home orchard, Transcendent is a good choice.

Tuna-apple salad

Unpared apples, as well as oranges and chopped pecans, make this tuna salad from **Marie Johnson** *of Tucson special.*

1 medium head lettuce, torn
   in bite-size pieces
2 cups diced unpared apples
1 orange, peeled and cut in small chunks
1 6½-ounce can tuna, drained and broken
   in chunks
⅓ cup coarsely chopped pecans

½ cup mayonnaise
2 tablespoons soy sauce
1 teaspoon lemon juice

Toss together in large bowl the lettuce, apples, orange, tuna and nuts. Combine mayonnaise, soy sauce and lemon juice and toss gently with salad. Serves 4 to 6.

Curried chicken with apples

Apples, other fruits and nuts make this chicken a special treat.

4 chicken breasts (boned, if desired)
½ cup flour
2 tablespoons oil
5 ounces chicken stock
½ teaspoon salt
Pepper to taste
1 clove garlic, minced
1 tablespoon dry sherry
1 tablespoon soy sauce
¾ teaspoon curry powder
1 tablespoon honey
1 tablespoon chopped parsley
2 medium apples, peeled and sliced
½ cup currants or raisins
½ cup broken pecans

Dredge chicken breasts in flour. Heat oil and fry chicken on high heat, 2 to 3 minutes, until brown. Remove from pan and discard oil. Cook together chicken stock, salt, pepper, garlic, sherry, soy sauce, curry powder, honey and parsley. Return chicken breasts and cook about 30 minutes. Add apples, currants or raisins and pecans. Cook about 10 minutes until chicken is done and sauce is thickened. Serves 4.

Sausage with apples and sweet potatoes

*Tucsonan **Nina Mitchell** has made this main-dish casserole many times. It comes from a recipe collection of Presbyterian women, and is a family favorite.*

½ pound sausage, bulk or links
2 medium apples
1 medium sweet potato
½ teaspoon salt
1 tablespoon flour
2 tablespoons sugar
½ cup cold water
1 tablespoon sausage drippings

Cut link sausage into ½-inch pieces; fry until well done. If bulk sausage is used, shape into small balls before frying. Pare and slice apples and potato. Mix salt, flour and sugar together; blend with water. Arrange layers of potatoes, apples and sausage in baking dish, pouring some of flour mixture over each layer. Top with apples and sausage; sprinkle with drippings. Cover and bake at 375 degrees until apples and potatoes are tender, about 45 minutes. Makes 3 to 4 servings.

Apple pancakes

*A delicious idea for breakfast or dessert is this German recipe from **Inge Jancic** of Tucson.*

1 cup flour
1 cup milk
¼ teaspoon salt
3 eggs, separated
3 medium apples, pared and sliced thin into rounds
Cinnamon and sugar
Shortening for frying

Combine flour, milk and salt to make a thin pancake batter. Add egg yolks and beat until well blended. Beat whites until stiff and fold into batter. Add apple slices and stir to coat. Have skillet very hot and use lots of shortening. Lift apple slices from batter, one at a time, and fry until brown on both sides. Sprinkle with sugar and cinnamon.

Honey applesauce bread

Pecans from the orchards near Green Valley make this recipe, used often by **Barbara Stockwell** *of Arivaca, into a really special bread.*

2 eggs
½ cup shortening
1 cup honey
2 cups flour
1 teaspoon soda
½ teaspoon salt
1 cup applesauce
1 cup chopped pecans

Cream honey, eggs and shortening. Beat until light. Mix in dry ingredients alternately with applesauce. Add nuts. Pour into greased loaf pan and bake for 60 minutes at 350 degrees. Makes 1 loaf.

Raw apple cake

Juanita McLennan *of San Manuel usually makes applesauce and pies for her freezer when she gathers the fruit from the family's two trees. But sometimes she likes to make this spicy cake.*

2 cups sugar
2 eggs, beaten
1 teaspoon salt
1 teaspoon cinnamon
3 cups diced or grated apples
1½ cups raisins
1½ cups oil
3 cups flour
1 teaspoon soda
1 teaspoon vanilla
1 cup chopped nuts
½ cup wheat germ

Mix all ingredients together. Flour and grease a tube pan. Bake at 325 degrees for 1 hour or until done. Turn out when just barely cool.

Sharon's apple coffeecake

Shirley Taylor *of Tucson developed this coffeecake for a farewell party for Sharon, an office staff-member, and it became a favorite dessert.*

Filling:
4 cups thinly sliced apples
1 tablespoon flour
⅔ cup sugar
1 teaspoon cinnamon
2 tablespoons water
2 tablespoons butter
1 tablespoon lemon juice

Batter:
2 cups flour
1¼ cups sugar
⅔ cup butter
2 teaspoons baking powder
1 teaspoon salt
2 eggs, separated
1 cup milk
1 teaspoon vanilla

To make filling, combine apples, flour, sugar, cinnamon, water, butter and lemon juice and cook until apples are tender.

To make batter, cut butter into flour and sugar into particles the size of coarse cornmeal. Reserve ¾ cup for topping. Add baking powder, salt, egg yolks, milk and vanilla and beat for 3 minutes on low speed of electric mixer. Beat egg whites until stiff and fold into batter. Pour in greased and floured 12-by-8-inch pan and spread filling evenly over the batter. Sprinkle with reserved topping mixture. Bake at 350 degrees for 40 to 50 minutes. Serves 10.

Tortilla apple strudel

*Flour tortillas are great as the base for apple strudel, **Hank Watchman** of Tucson has found. When you make the filling, you can use your own variation on this guide, he says. (Add a little rum, for instance.)*

1 package large flour tortillas
2 quarts cooking apples, pared and cut fine
1 cup raisins
4 ounces nuts, chopped
¾ cup sugar
2 teaspoons cinnamon
½ cup melted butter
½ cup corn flakes, crushed

Mix apples, nuts and raisins. Stir in sugar mixed with cinnamon. Place tortilla on flat surface and spread with melted butter, using brush. Sprinkle some crushed cornflakes lightly over buttered tortilla and then spread with some of apple mixture.

Fold ⅓ of tortilla over mixture, brush with butter, fold both sides of tortilla to center, brush with melted butter and then roll. Brush entire outside with melted butter. Repeat for each tortilla. Place rolled tortillas on buttered baking pan and bake at 350 degrees until brown and crisp. They can be frozen after baking.

Apples with applejack sauce

Try this interesting poached apple dish, spiced and flamed with applejack.

4 firm, slightly tart apples
3 tablespoons butter
½ teaspoon grated lemon rind
4 tablespoons sugar
¼ teaspoon cinnamon
¼ cup applejack
4 scoops vanilla ice cream

Peel and quarter apples and remove cores. Cut quarters in thick slices.

Heat the butter in a skillet and add the apple slices and lemon rind. Sprinkle with sugar and cinnamon and cook, stirring the apples and shaking the skillet so that the apples cook evenly. Cook over high heat. When the apples start to brown, add the applejack and ignite it. Blend well.

Serve apples and sauce on top of ice cream in each of 4 individual serving dishes.

Good apple dessert

***Alberta Fienhold** of Tucson receives compliments when she serves this good apple dessert.*

1 cup brown sugar
¼ cup butter
1 cup flour
½ teaspoon nutmeg
½ cup nuts
2 cups apples, diced
1 egg, beaten
1 teaspoon soda
½ teaspoon cinnamon
¼ teaspoon salt

Combine ingredients and mix with hands as batter is stiff. Put in greased and floured square pan and bake at 325 degrees for 45 minutes. Serve topped with hot brown-sugar sauce.

Sauce: Combine and cook for 5 minutes, stirring: ¼ cup each butter, white sugar and brown sugar, 1 teaspoon vanilla and ¼ cup cream or milk.

Apples

Cocoa apple-nut cake

This great apple cake goes to Tucsonan **Mary Ellen Condon's** *son in San Francisco as a "CARE" package, and understandably draws much gratitude.*

2 cups granulated sugar
3 eggs
1 cup margarine
½ cup water
2½ cups all-purpose flour
2 tablespoons cocoa
1 teaspoon each baking soda, cinnamon and allspice
1 cup finely chopped nuts
½ cup semisweet chocolate bits
2 cups grated apples
1 teaspoon vanilla

Cream the eggs, sugar, margarine and ½ cup water until fluffy. Sift together the flour, cocoa, soda, cinnamon and allspice and add to creamed mixture. Mix well. Fold in the nuts, chocolate bits, grated apples and vanilla and mix well. Spoon into greased and floured 10-inch spring-form tube pan. Bake at 325 degrees for 60 to 70 minutes or until cake tests done. Cool upside down for 10 to 15 minutes before removing from pan. Serves 12.

Apple crisp

Janie Haller *shares this apple crisp, which she makes with some of Thatcher's good fruit.*

4 cups sliced, pared and cored baking apples (about 6)
¾ cup brown sugar, packed
½ cup flour
¾ teaspoon cinnamon
⅓ cup soft butter
½ cup rolled oats

Place sliced apples in greased square 8-by-8-inch pan or 1½-quart baking dish. Combine remaining ingredients until mixture is crumbly. Spread over apples and bake for 35 minutes at 375 degrees or until apples are tender and topping browned. Serve warm with ice cream. Serves 8.

Knibb apple pie

Muriel E. Hollis *of Tucson says this recipe is a family favorite and comes from her great-grandmother Frances Bowles Knibb of Virginia. It's more of a cobbler than a pie.*

Pastry for 4, 9-inch pie crusts
6 small to medium-size sweet potatoes, boiled and peeled
6 tart apples, peeled
Brown sugar, butter, ground cloves, nutmeg
1½ cups seedless damson plum or seedless grape preserves
1 quart pickled pears or spiced peaches
¾ cup blackberry wine (or hot water)
Heavy cream

Line a deep 9-by-12-inch baking with a fairly thick pie crust. Cover with layer of ¼-inch sweet potato slices, then a layer of thinly sliced apples. Sift brown sugar over apples, dot with butter and sprinkle with cloves and nutmeg. Next, put on a layer of preserves and a layer of pickled pears cut in small pieces. Repeat until dish is filled, but don't crowd. Pour the wine or hot water over the mixture, adding additional hot water as needed, but do not allow to become too juicy. Cover with top crust, slash generously and bake at 350 degrees for about 1 hour, until crust is brown. Serve in deep dishes with thick cream. Serves 12.

Cobbler for a crowd

Sabina Larson *of Bonita, whose cooking skills are challenged when she prepares meals for the cowboys at round-up time on the Larson ranch, passes along these tips for success:*

You can make the pastry in a jiffy with biscuit mix, which you enhance with a little brown sugar, some cinnamon and nutmeg and lots of vegetable oil to give it richness. Spice the apples with more brown sugar, cinnamon and nutmeg. And, to be sure your apples will be done when the crust is done, precook them a little.

Butterscotch apples

Lucia Eppinga *of Tucson devised this recipe years ago and still enjoys making it.*

2 cups favorite butterscotch pudding
1 cup soft bread crumbs
2 cups applesauce

Combine bread crumbs with pudding. Place a cup of applesauce in buttered baking dish and cover with half the crumb-pudding mixture. Add another cup of applesauce and cover with remaining crumb-pudding mixture. Cover dish and bake at 350 degrees for about 1 hour.

Apple goodie

As a working woman, **Arleta Snyder** *of Willcox finds this jiffy recipe much appreciated.*

3 cups sliced apples
1 scant cup sugar
1 rounded tablespoon flour
    Pinch of salt
    Sprinkling of cinnamon
¼ teaspoon soda
¼ teaspoon baking powder
⅓ cup melted butter
¾ cup oatmeal
¾ cup flour
¾ cup brown sugar
    Hard sauce or whipped cream

Combine apples, sugar, flour, salt and cinnamon in a greased 8-inch square baking dish. Mix remaining ingredients and top fruit. Bake at 425 degrees for 30 minutes. Serve with hard sauce or whipped cream.

Applesauce cookies

Florenia Allen *of Safford shares this cookie recipe from her mother,* **Minnie Watson,** *also of Safford, who is now in her 90s.*

1 cup applesauce
1 cup raisins
1 cup sugar
½ cup shortening
1 egg
2 cups flour
1 teaspooon each salt, baking powder,
    cinnamon, nutmeg and cloves
½ teaspoon soda
1 cup nuts

Mix applesauce and raisins together and set aside. Cream shortening and sugar, add egg and beat well. Stir in applesauce and raisins. Sift flour, salt, baking powder, spices and soda and mix well. Add nuts. Drop by the teaspoonful on cookie sheet. Bake at 375 degrees for 12 to 15 minutes.

Quick apple relish

Some lemon and a little mustard give plenty of oomph to this no-cook relish that will keep several weeks in the refrigerator.

1 tablespoon Dijon-type mustard
3 tablespoons olive oil
¼ teaspoon paprika
½ teaspoon salt
1 tablespoon sugar, or to taste
Juice of 1 medium (3 ounces) lemon
1½ teaspoons cider vinegar
3 to 4 tart apples, peeled and grated

In a mixing bowl, gradually beat olive oil into mustard, using a fork or wire whisk. Stir in remaining ingredients. Taste and adjust seasoning. Serve at room temperature. However, recipe must be kept refrigerated in a tightly covered container if there are leftovers. Flavor is best when used within a month. Makes about 2 cups.

Best apple butter

*Refrigerate this apple butter for safe-keeping, suggests Tucsonan **Virginia E. Bach**. However, it's so good it won't last long.*

3 quarts plain applesauce
4 cups sugar
¼ teaspoon ground cloves
¼ teaspoon allspice
1 teaspoon cinnamon
½ cup cinnamon "red hot" candies
12 whole cloves
6 cinnamon sticks

Heat applesauce and add sugar and spices, except for whole cloves and cinnamon sticks. Carefully stir often, especially until the red hots are dissolved. Cook and stir on low heat until applesauce thickens and has a glazed look. Process takes about 2 hours. (Use a wooden slotted spoon to stir with.) Transfer apple butter to clean glass jars, stir in one cinnamon stick and 2 cloves in each and fasten lids. Makes 6 pints.

Crab apple jelly

Among the favorite uses for the little crab apple is making jelly. This version includes lemon peel.

4 pounds crab apples
Water
Peel of 2 lemons
Sugar

Cut crab apples in quarters and place in kettle with water to cover. Add lemon peel and bring to a boil. Simmer and stir occasionally until apples are tender, about 30 minutes. Strain through a jelly bag, measure the juice and add 1¾ cups sugar to each 2 cups juice. Bring to boil, and boil for about 10 minutes or until the juice tests for jelly (sheets from a metal spoon). Remove from heat, skim off foam and pour into sterilized jelly glasses and seal. Makes 7 or 8 6-ounce glasses.

Spiced crab apples

*Tucsonan **Hanna Lundberg**'s recipe is based on one she got from the Kerr Glass Manufacturing Corp. exhibit at the Chicago World's Fair in 1934.*

3 pounds tart crab apples
4 cups sugar
1 cup vinegar (dark or light)
1 cup water
2 sticks cinnamon
2 tablespoons whole cloves

Select fruit of uniform size, wash and cut off the blossom ends. Using a large needle, prick deeply through the blossom end so fruit won't burst. Leave stems on. Prepare a syrup by boiling the sugar, vinegar, water and spices together for 5 minutes. Pour over apples and cook until tender. Let fruit stand in syrup overnight. Reheat and pack in jars.

Apple-tomato barbecue relish

Tucsonan **Mary A. Coulter** *can be sure of the compliments when she makes this barbecue relish.*

12 apples
12 tomatoes
12 onions
2 hot peppers
2 green peppers
1 pint vinegar
2 cups sugar
1½ teaspoon salt
1 teaspoon dry mustard
1½ teaspoons celery seeds

Peel, quarter and core the apples; peel and cut onions in quarters and quarter the tomatoes. Seed the peppers. Grind these together and transfer to large kettle. Add vinegar, sugar, salt, dry mustard and celery seed and cook for 50 minutes. Pack in jars and seal. Makes 4 or more pints.

Apple chutney

An old-fashioned chutney is spicy and delicious and a lot easier to make than most people realize.

10 cups cooking apples, peeled,
  cored and sliced (about 5 pounds)
4 cups firmly packed light brown sugar
4 cups cider vinegar
2¼ cups raisins
1 cup chopped onions
3 tablespoons mustard seed
1 tablespoon ground ginger
2 teaspoons ground allspice
¼ teaspoon garlic powder
  Dash ground red pepper

In large stainless steel or enamel kettle combine apples, brown sugar, vinegar, raisins, onions, mustard seed, ginger, allspice, garlic and red pepper. Bring to a boil. Reduce heat and simmer, covered, stirring frequently until thickened, about 1¼ hours. Pour into 6 1-pint clean, hot canning jars, leaving ¼ inch head space. Cover, following manufacturer's directions. Process in boiling water bath for 20 minutes. Cool jars and check seals. Makes 6 1-pint jars.

Apricots

Botanical name: *Prunus armeniaca*

Apricots are a nutritionist's delight. They are high in vitamin A, and contain a good supply of other vitamins and minerals.

And what a savory treat it is to eat lush, ripe apricots from the tree. Luckily, the enjoyment can be extended far beyond their short May-June season by canning, preserving or freezing. They are among the most popular of fruits for drying.

Apricots had their beginnings in China, and are justly famed as an important food on the tables of the long-lived people of Hunza, high in the Himalayas. Alexander the Great is credited with introducing the fruit to Europe in the 4th century B.C., and it was popular in Rome in Jesus' time. Eventually, apricots arrived in England where they thrived under the careful attention of Henry VIII's gardeners. The tree was growing in Virginia by the early 1700s.

Apricots came to Arizona by way of California, which still provides most of the supply sold at the supermarkets.

For the yard, wide-spreading apricot trees furnish excellent shade, and their pink blossoms add spring charm.

In low-chilling areas, Royal, Blenheim and Katy are good early apricots. Maricopa County finds the early ripening Patterson and Modesto successful. Caselton does well in the Tucson area. In some of the cooler areas of Maricopa and Pinal counties and at 2,000 to 3,500-foot elevation, the later-ripening favorites are Reeves and Tilton. From 3,500 to 6,000 feet, Perfection is a good apricot for canning; Wenatchee and Goldrich are later apricots that offer good eating.

Ham and apricot kebabs

Fresh apricots and other fruits are combined for this delectable main dish with ham.

3 pounds cooked ham, cut into 1-inch cubes
6 apricots, halved and seeded
3 apples, cut in wedges
3 green bananas, cut into 1½-inch pieces
1 cup grape jam
2 tablespoons honey
1¼ teaspoon curry powder
⅛ teaspoon ground ginger

Arrange ham, apples, apricots and bananas on skewers. In a small saucepan, combine jam, honey, curry and ginger. Heat, while stirring, until jam is melted and sauce is heated through.

Grill kebabs 3 to 5 inches from heat, basting with grape sauce. Turn and baste several times until heated through. Serve hot on cooked rice. Serves 6.

Frosted apricot salad

Henrietta Brooks of Tucson is sure to make this salad whenever her sons come home for a visit.

2 3-ounce packages orange gelatin
4 cups boiling water
1 No. 2 can crushed pineapple, drained
2 cups canned apricots, quartered and drained
¾ cup miniature marshmallows
1 cup liquid drained from fruits
½ cup sugar
1 egg
3 tablespoons flour
1 teaspoon butter
½ pint whipping cream

Dissolve gelatin in boiling water and chill until partially set. Fold in pineapple, apricots and marshmallows and return mixture to refrigerator to chill. Combine 1 cup liquid from fruits, sugar, egg and flour in saucepan. Cook and stir until thickened. Add butter and blend well. Let cool and then fold in the whipped cream. Spread over the congealed salad. Serves 10 or more. Recipe may be cut in half for family use.

Apricot nut bread

This quick bread has been a competition winner for **Maxine Haverstick** *of Tucson.*

    2 cups dried apricots, plumped in water
    4 tablespoons shortening
    2 cups sugar
    2 eggs
    ½ cup apricot liquid
    1 cup orange juice
    4 cups flour, sifted
    4 teaspoons baking powder
    ½ teaspoon soda
    1 teaspoon salt
    ½ cup nuts, chopped

Cut apricots in ¼-inch pieces and add water to plump. Drain and save liquid for recipe. Cream shortening, sugar and eggs. Stir in ½ cup apricot liquid and orange juice. Sift dry ingredients together and stir into shortening mixture. Blend in nuts and apricots.

Pour in 2 loaf pans, 9-by-5-by-3 inches, that have been lined with wax paper. Let stand for 20 minutes. Bake for 55 to 65 minutes at 350 degrees. Cool for 20 minutes; turn out of pans and peel off paper. Turn 2 loaves right side up on rack. Bread freezes well.

Apricot leather layered bread

One of the favorite breads of solar-cooking enthusiast **George Brookbank** *of Tucson is made with apricot leather (or use peach).*

    1 heaping tablespoon active dry yeast
    ½ cup warm water
        All-purpose flour
    ¼ cup honey
    2 cups whole-wheat flour
    1 cup non-fat dry milk powder
    ½ cup softened margarine
    1 egg
        Apricot fruit leather

Dissolve yeast in water and stir in ½ cup flour and honey. Combine whole-wheat flour, non-fat dry milk, margarine and egg and then work into the yeast mixture to form a stiff dough, adding more all-purpose flour as needed. Knead, and roll flat with a rolling pin to about ½-inch thick. Cut into 4 or 5 pieces to fit into a 9-by-4½-by-5-inch loaf pan. Place first layer in bottom of loaf pan and top with a ½-inch layer of dried apricot leather. Repeat the layers, ending with the dough. Let rise in warm place, then bake in solar oven until bread is done and top is browned. (If baked in conventional oven, bake at 375 degrees until done and top is browned.) Top with orange glaze: ½ cup powdered sugar, 1½ teaspoons grated orange peel and orange juice to moisten.

Apricot-date-nut bread

Great for doing in your solar oven; but also can be baked in a regular oven, says Tucsonan **Syd Clayton**.

    2 cups whole wheat flour
    1 teaspooon baking soda
    1 teaspoon baking powder
    1 egg
    1 tablespoon butter
    ⅔ cup brown sugar
    1 cup orange or apricot juice
    6 to 8 apricots, peeled and mashed (or use
        1 cup dried, cut up)
    1 cup pitted and chopped dates
    1 cup chopped nuts
    ½ cup powdered milk

Combine ingredients and put in greased loaf pan. Bake at 350 degrees for 45 to 50 minutes, in conventional or solar oven.

Apricot

Gluten-free apricot tarts

*For those on gluten-free diets, **Marion N. Wood** of Globe sends along this tasty dessert by special permission of Charles C. Thomas, publisher of "Coping With the Gluten-Free Diet."*

Apricot jam for filling
1½ cups brown rice flour
⅓ cup tapioca starch or cornstarch
½ teaspoon salt
½ cup butter
4 ounces cream cheese
1 egg
1 tablespoon milk (about)

Cut shortening into sifted dry ingredients; cut in cream cheese; add slightly beaten egg; sprinkle the milk over the pastry mixture and stir and knead into a ball. Roll out pastry between 2 sheets of plastic wrap. Mark top with 3-inch fluted tart-pan shape. Peel off top plastic sheet and cut out tarts, using the edge of the tart pan to cut. Transfer pastry to pan and smooth out. Prick pastry all over. Bake at 400 degrees for about 10 minutes or until lightly browned. Fill each tart with jam and top with daub of whipped cream, if desired. Also good with raspberry or strawberry jam.

Apricot-glazed tarts

*These pretty tarts, suggested by **Mary A. Hardy** of Tucson, are topped with apricots, orange, plums, peaches, etc., and glazed with apricot preserves for a delicious party treat.*

Crust:
1½ cups flour
¼ cup sugar
¾ cup butter or margarine

Cheese filling:
12 ounces cream cheese
⅔ cup sugar
1½ tablespoons lemon juice
3 eggs

Glaze:
¾ cup apricot preserves
3 tablespoons lemon juice

To make crust, mix flour with sugar; add softened butter and mix well. Press into greased individual 2-inch tart pans or 10-inch pie plate.

To make filling, beat cream cheese in an electric mixer until soft. Add sugar, lemon juice and eggs and beat well. Pour into small tart shells. Bake for 25 to 30 minutes at 350 degrees until a knife inserted in center comes out clean. Cool.

Heat preserves and lemon juice slowly to boiling. Arrange sliced fruits on baked tarts. Pour hot glaze over fruit. Makes 1½ dozen small tarts or 1 large pie.

Peppy popcorn

Pep up popcorn with dried apricots and nuts this way: Pop 2½ quarts popcorn. Then pour ¼ cup melted butter over the hot popcorn and toss to coat. Next, toss with 1 cup wheat germ, ½ cup sunflower seeds and ½ cup chopped dried fruits.

Apricot filling

Elizabeth Ferguson *of San Manuel dries the apricots from her tree to use in many ways, including this tasty cake filling.*

2 cups dried apricots
1½ cups sugar
1 tablespoon lemon juice
½ teaspoon grated lemon rind

Cover apricots with water and cook until tender and very little juice remains. Drain off remaining juice and set aside for frosting. Beat apricots 1 minute at medium speed of mixer to thoroughly mash; add sugar, lemon juice and rind and continue beating ½ minute to dissolve. Return to heat and cook for 1 minute. Cool and use as filling for cake. Makes enough filling for 2 layers of a 3-layer cake. Make confectioners' sugar frosting for top of cake, using reserved apricot liquid as needed.

Cordial fruits

Serve these fruits with cheese. The liquid, chilled, makes a cordial.

2 cups (½ pound) dried apricots
2 cups (½ pound) dried apples
2 cups (½ pound) dried pears
1 bottle medium dry white wine
1 cup brandy
¼ cup honey
½ cup brown sugar
1 stick cinnamon

Cover apples with water and simmer for 5 minutes. Drain. Combine apples with apricots and pears in large jar. Combine wine, brandy, honey, brown sugar and cinnamon. Stir until sugar is dissolved. Pour over fruits; cover and let stand 1 week or longer in cool place. Makes 7 cups fruit and 2¾ cups liquid.

Note: Additional fruit may be added to the wine-syrup after the first prepared fruits are used.

Apricot pot de creme

Dried apricots make this dessert a special dish that is also special in nutrition: One serving contains a good portion of an adult's daily needs of vitamin A.

8 ounces dried apricots
1 cup plus 2 tablespoons water
⅓ cup honey
1 tablespoon lemon juice
2 tablespoons apricot-flavored brandy
1 cup heavy cream, whipped

Simmer apricots in a covered saucepan with water, honey and lemon juice until fruit is tender. (Add more water if needed.) Cool to room temperature. Stir in apricot brandy. Purée until very smooth in blender or food processor with steel blade. Pour apricots into a bowl and gently fold in whipped cream. Spoon into 6 pot de creme dishes, sherbet cups or wine glasses. Refrigerate for several hours. Garnish with whipped cream. Serves 6.

Fried apricot pies

Dried apricots or other dried fruits are just right for this fried pie recipe from **Beth Oldfather Gladden** *of Marana.*

Filling:
Simmer dried fruit in enough water to reconstitute. Add sugar and cinnamon to taste and cool. You will need about 1¼ cups filling.

Pastry:
2 cups flour
2 teaspooons baking powder
1 teaspoon salt
2 tablespoons shortening
Water
Oil for frying
Powdered sugar

Sift flour, baking powder and salt and

cut in shortening to consistency of coarse cornmeal. Add water to make a ball. Let stand for 10 minutes before rolling out into a thin sheet. Cut into 8 squares. Top with rounded tablespoon of filling and fold pies over into triangles. Pinch edges together and let stand for 10 minutes. Deep fat fry at 375 degrees, turning once. Drain and sprinkle with powdered sugar. Serves 8.

Apricot bars

Dried apricots are the fruit that brighten the flavor of these bar cookies from Tucsonan **Kay Trondsen.**

⅔ cup dried apricots
½ cup soft butter or margarine
¼ cup granulated sugar
1⅓ cups sifted flour
½ teaspoon baking powder
¼ teaspoon salt
1 cup brown sugar
2 eggs, well beaten
½ teaspoon vanilla
½ cup chopped nuts
   Confectioners' sugar

Rinse apricots and cover with water. Boil for 10 minutes. Drain, cool and chop. Combine butter, granulated sugar and 1 cup flour until crumbly. Pack into greased 8-by-8-by-2-inch pan. Bake for 25 minutes.

Sift ⅓ cup flour, the baking powder and salt. Gradually beat brown sugar into eggs. Mix in flour mixture, then vanilla and nuts and apricots.

Spread over baked layer. Bake for about 30 minutes or until done. Cool in pan. Cut into 32 bars; roll in confectioners' sugar.

Apricot-granola fruit bars

Martha Schuetz of Tucson modified a fig bar to make use of a gift of fresh apricots.

2 cups chopped fresh apricots
¼ to ½ cup water
3 tablespoons honey or ½ cup sugar
¾ teaspoon almond extract
1 cup brown sugar
½ cup butter or margarine
1¼ cups unbleached flour
1¼ cups regular rolled oats, uncooked
½ cup chopped pecans
¼ cup sesame seeds
½ teaspoon each salt and soda

To prepare filling, combine apricots with water and simmer until apricots fall apart and mixture thickens. Mix in honey and almond extract.

Cream brown sugar and butter. Mix together the flour, oats, nuts, seeds, salt and soda. Slowly stir into sugar mixture until well-blended. Press half the mixture into a greased 8-inch square pan. Spread with fruit filling. Spread or crumble remaining oat mixture over top, pressing lightly into filling. Bake at 350 degrees for 25 minutes or until golden. Makes 16 bars.

Apricot nut candy

1 cup dried apricots
¾ cup shredded coconut
½ cup pecans or almonds
¼ cup coconut milk
   Grated coconut or finely chopped nuts

Finely grind apricots, shredded coconut and nutmeats. Mix well with coconut milk. Shape into balls the size of walnuts and roll in grated coconut or finely chopped nuts. Store, covered, in the refrigerator. Makes about 2½ dozen balls.

Apricot fruit balls

Dried apricots combine with orange for this appetizing confection.

1 pound dried apricots
1 seedless orange
2 cups sugar
  Granulated sugar

Rinse dried apricots and dry them on paper towels. Peel and cut the seedless orange in pieces and put apricots and oranges through a grinder or blend coarsely. Place in top of a double boiler with the 2 cups sugar and steam until sugar is dissolved. Cool and shape into 3½-dozen balls. Roll in sugar.

Apricot-almond conserve

When fresh apricots are on the tree, make this conserve to enjoy the fruit all year long.

3 pounds fresh apricots
2 oranges
4 cups sugar
2 teaspoons ground ginger
½ cup slivered almonds

Wash, halve and pit apricots. Peel oranges and cut into eighths. Put apricots and oranges through medium blade of food grinder. Place in large kettle. Add sugar and ginger and bring to a boil. Simmer until thick, about 1 hour. Stir in almonds. Ladle into sterilized jars and seal. Process for 10 minutes in boiling water bath. Makes about 8 ½-pints.

Cherries

Botanical name: *Prunus avium (sweet cherry); P. serotina (wild)*

Orchard-type cherries like temperate climates, and thus are not associated with desert cultivation. However, in cooler areas of counties adjoining the Sonoran Desert, from 3,500 to 7,000 feet, home growers find that Bing and Van (smaller than Bing) both make good eating. Rainier and Lambert are other varieties recommended. Except for Rainier, which is a large, pink cherry, these varieties are dark fleshed. All are sweet cherries requiring a pollinator. They fruit in summer.

Nutritionally, sweet cherries are no blockbuster. They contain small amounts of vitamins A and C and fair amounts of potassium, but their eating pleasure ranks very high, whether fresh or canned. They also can be frozen.

Wild, or chokecherries, are found at elevations above 4,000 feet, in pine country. Their tiny white blossoms are fragrant, and their small fruits, which ripen in July and August, are dark and sour, and are preferred in jam or jelly to eating fresh.

Chokecherry jam

Hikers who find chokecherries can make interesting jam with this old-style recipe.

4 cups berries
1 cup water
Sugar

Wash and remove stems of berries. Place in saucepan with water and bring to simmer, stirring occasionally until tender. Put berries through a medium sieve and measure. Add 1 cup sugar to each cup pulp. Cook and stir over moderate heat until sugar dissolves; then cook and stir on high heat until mixture reaches a rolling boil. Continue to cook and stir until jam sheets from a spoon. Seal in hot sterilized jars. For those who like almond flavoring, a few drops may be stirred into the jam. Recipe makes 3 ½-pints.

Cherry jam

Sweet cherries make an interesting refrigerator jam.

1 quart pitted and coarsely
  chopped sweet cherries
¼ cup sugar
¼ cup cream sherry
1 envelope unflavored gelatin
⅓ cup cold water

Combine cherries, sugar and sherry in saucepan and heat for 5 minutes, stirring occasionally. Bring to boil and boil rapidly for 3 minutes. Soften gelatin in cold water. Add to cherry mixture and heat for about 3 minutes or until gelatin dissolves. Let stand for 5 minutes; skim off foam. Pour into freezer-proof containers, cover and let cool. Store in refrigerator for 1 to 2 weeks or freeze up to a month. Makes 3 cups.

Freezing cherries

Add ⅓ cup sugar to each pint of pitted or unpitted sweet cherries. Fill freezer containers, shaking container to pack cherries closely, leaving ½-inch headroom. Cover tightly and freeze.

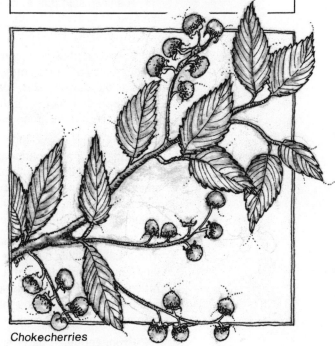

Chokecherries

Cherry salad

When fresh sweet cherries are available, combine them with other fruits and nuts for a salad.

1½ cups sweet cherries
2 large apples
¼ cup pecans
  Salad dressing (Miracle Whip or similar)
½ cup heavy cream, whipped stiff
  Fresh mint leaves

Pit cherries, drain and save juice. Chill fruit. Pare and core apples, cut in eighths, then slice crosswise in ½-inch pieces. Combine with cherries. Add nuts and a spoonful of dressing. Toss and coat lightly, adding a little more dressing to coat well. Add a little cherry juice to whipped cream and spoon onto salad. Chill. Serves 4.

Dates

Botanical name: *Phoenix dactylifera (edible dates).*

It's been said that a date palm "must have its feet in water and its head in the fires of heaven." In our sunny desert, wherever water is available, these needs are easily filled.

Stately, edible date palms, some as tall as 60 feet, provide delicious fruit, although they are more often used as ornamentals. For show only, there are such types as the Canary Island date palms, which do well throughout the desert, and the Senegal date palm and the pygmy date palm, both of which prefer the warmer areas of the Sonoran Desert.

The Canary Island date palm's trunk is pineapple-shaped when young, becoming thick and columnar at maturity. The tree can grow to about 50 feet, and is topped with a fountainlike, wide-spreading crown of green fronds. The smaller Sengal palm develops curved multitrunks. The pygmy palm is ideal for container culture. All three of these ornamentals produce small fruit, mostly skin and stones, of immense delight to birds.

The edible date palm may be identified by its straight, slender trunk topped with a tall-standing head of gray-green leaves, somewhat lacier than the leaves of the Canary Island date palm. Treating the edible date palm as an ornamental rather than a fruit tree has some justification: Handling the fruit is difficult. Ladders must be used to reach the fruit, for thinning, hand pollinating, etc. The palms may begin bearing fruit as early as seven years after planting.

From antiquity, dates have furnished sweet, nourishing food to desert peoples. Date palms may have flourished in the Garden of Eden, and the Babylonians likely cultivated them as long as 8,000 years ago.

Dates are still an important factor in the economy of Arab countries; they grow well also in Spain, North Africa, Southwest Asia, India and Pakistan. The Spanish brought them to the New World, where they thrived in the desert areas of both Arizona and California.

Today, dates are a crop of much value in California, but few are grown commercially in Arizona. At the end of World War II, for example, date gardens were surrounded by the city of Phoenix and were sold for home sites.

The yield and food value of dates — one-seeded fruits known as *drupes* — make it well worth the time it takes to handle them. A medium-size tree can provide dozens of pounds of dates in a season. As for their nutrients, dates are low in sodium and high in potassium, and contain a fair amount of B vitamins and vitamin A, but no vitamin C. Very sweet and flavorful, dates are often used to enhance cookies and cakes.

Depending on variety, dates are tannish to reddish to very dark when ripe and can be eaten fresh. Most people like to give them a light heat treatment to destroy possible contamination. (See cleaning instructions at end of date section.)

Unless this treatment is followed by further dehydration or by canning under pressure, the dates should then be refrigerated. Fully cured dates have a sugar content of about 50 percent, and their calorie count is comparable to dried peaches and figs.

Among popular varieties for home orchards are Halawy and Khadrawy (Iraqi varieties) and Deglet Noor (Algerian variety).

Sourdough date loaf

Tucsonan **Virginia B. Selby** *shares this date-loaf recipe for which a sourdough starter is required.*

½ cup sourdough starter
1½ cups flour
1 cup undiluted evaporated milk
2 tablespoons sugar
¼ cup butter or margarine
¾ cup brown sugar, firmly packed
1 cup chopped dates
½ cup chopped nuts
2 eggs, beaten
½ cup quick-cooking rolled oats
1 teaspoon baking powder
½ teaspoon each baking soda and salt

The night before baking is planned, combine starter, flour, evaporated milk and sugar; cover and leave at room temperature overnight. The next day, cream butter and brown sugar. Add dates and nuts; set aside. Combine eggs, rolled oats, baking powder, soda and salt; stir into sourdough mixture with date mixture. Turn into a greased 5-by-9-inch loaf pan and let rise about an hour. Bake at 375 degrees for an hour. Cool for 10 minutes in pan, then remove from pan to cooling rack. Serve warm or cool.

Date bread

Etta Mattausch *of Phoenix makes this tender, moist bread recipe with dates from the family's date tree.*

1 cup pitted dates
1 cup raisins
2 cups water
2 tablespoons shortening or margarine
1 cup sugar
2¾ cups flour
2 teaspoons baking soda
½ teaspoon salt
1 teaspoon cinnamon
1 egg, beaten

Boil dates and raisins with water for 2 or 3 minutes. Cool. Mix in remaining ingredients to make well-blended dough. Bake in greased and floured 9½-by-5¼-inch bread pan at 350 degrees until done, about 1 hour. Makes 1 loaf.

Date-cheese bread

Peggy Putnam *of Casa Grande highly recommends this flavorful bread.*

¾ cup boiling water
½ pound (1¼ cups) dates, finely chopped
1¾ cups sifted flour
1 teaspoon baking soda
½ teaspoon salt
½ cup sugar
1 egg, well beaten
1 cup shredded Cheddar cheese
1 teaspoon vanilla
¾ cup chopped pecans

Pour water over dates and let stand for 5 minutes. Sift together the flour, soda, salt and sugar. Add date mixture, egg, cheese, vanilla and nuts; mix well. Pour into greased 9-by-5-by-3-inch loaf pan. Bake at 350 degrees for 45 to 50 minutes. Remove from pan. Cool on rack. Makes 1 loaf. Bread freezes well.

Kahlua bread

Though she doesn't usually eat desserts, Tucsonan **Betty Birkett** *finds this bread can be used as one. The idea was sent to her from Middletown, Ohio.*

1 cup chopped dates
½ cup coffee liqueur
½ cup warm water
1 teaspoon grated orange rind
⅔ cup packed brown sugar
2 tablespoons shortening
1 large egg
2 cups sifted all-purpose flour
1 teaspoon soda
1 teaspoon salt
⅔ cup chopped pecans

Combine dates, liqueur, water and orange rind. Let stand while preparing batter. Beat sugar, shortening and egg together until fluffy. Sift flour with soda and salt. Add to creamed mixture alternately with date mixture. Stir in pecans. Turn into greased loaf pan. Let stand for 5 minutes, then bake below oven center at 350 degrees for 60 to 70 minutes or until loaf tests done. Turn out on wire rack to cool. Serve sliced and spread with cream cheese spread.

Cream cheese spread: Beat 8 ounces of cream cheese, softened, with ¼ cup softened butter. Stir in 1 tablespoon coffee liqueur.

Date-onion relish

This unusual recipe was given to **Josephine Block** *of Chandler years ago by a woman who had served in the Peace Corps in North Africa.*

Pour boiling water over 1 good-sized onion, chopped, and let stand for 10 minutes. Drain well. Combine ¼ cup red wine vinegar, ¼ cup red wine, 1 tablespoon sugar and ½ teaspoon salt. Place 1 pound finely chopped dates and the onion in layers in small container, beginning and ending with dates. Pour wine mixture over dates and onions and let stand 2 or 3 days before using. Keep refrigerated. Will keep a minimum of 3 weeks.

Holiday torte

A must at Thanksgiving and Christmas in her family for many years has been Tucsonan **Gayle Pittman's** *cranberry and date torte.*

2¼ cups flour
¼ teaspoon salt
1 teaspoon baking powder
1 teaspoon baking soda
1 cup sugar
2 eggs, beaten
1 cup buttermilk
¾ cup vegetable oil
1 cup cranberries, cut up
1 cup diced dates
1 cup chopped nuts
Pineapple glaze (see below)

In a bowl sift together the flour, salt, baking powder, baking soda and sugar. Set aside. Combine the eggs, buttermilk and oil and mix in the cranberries, dates and nuts. Add the flour mixture and blend well. Pour into greased tube pan. Bake at 350 degrees for 1 hour. **Cool in oven;** remove when lukewarm.

Place on rack over large dish. Pour pineapple mixture over cake. Repeat until mixture is used up. Wrap cake in foil. Refrigerate for 24 to 48 hours before slicing.

Glaze: Combine 1 cup each pineapple juice and sugar.

Date torte

This date dessert from **Ellen Kightlinger** *of Tucson is as simple to make as it is delicious.*

3 egg whites
¼ teaspoon cream of tartar
1 cup sugar
2 tablespoons flour
1 cup dates, chopped
½ cup pecans, chopped
¼ teaspoon vanilla

Beat whites until stiff, but not dry. Sift together the cream of tartar, sugar and flour and add to the egg whites. Add dates, pecans and vanilla. Pour into greased baking pan 9-by-13-inch pan. Bake at 300 degrees for 30 to 35 minutes or until torte tests done. Cut in squares and serve topped with ice cream.

Dates

Date-zucchini torte

This dessert was used for a class **Jolene Randall** *of Tucson taught on "What to do with your summer vegetables."*

1 cup dates, diced
2 cups zucchini, grated
½ cup chopped nuts
1 cup brown sugar
2 eggs, beaten
1 tablespoon oil
1 teaspoon vanilla
½ teaspoon salt
1 teaspoon cinnamon
¾ cup whole wheat flour
2½ teaspoons baking powder

Beat eggs, brown sugar, oil, vanilla, salt and cinnamon in a bowl. Slowly stir in combined flour and baking powder. Stir in dates, zucchini and nuts. Pour into a greased 9-by-9-inch pan. Bake at 350 degrees for 1½ hours. Let sit for 1 hour before serving. Cut into squares. Serve warm or cold with whipped cream or vanilla ice cream.

Ingram eggless cake

Helen Ingram *of Tucson shares this old-fashioned cake that holds together without eggs.*

1 cup brown sugar
1 cup water
2 cups chopped dates
⅓ cup shortening
1 tablespoon cocoa (optional)
1 teaspoon each cinnamon and allspice
½ cup cold water
1 teaspoon baking soda
2 cups flour
1 teaspoon baking powder
¼ cup chopped pecans

In a saucepan, mix together the brown sugar, 1 cup water, dates, cocoa, shortening, cinnamon and allspice. Stir and bring to a boil. Cool. Dissolve soda in ½ cup cold water, and add to first mixture. Stir in flour, baking powder and pecans. Bake at 325 degrees for at least an hour in greased cake pan. Test for doneness with toothpick.

Rich Christmas fruitcake

*Dates, raisins, candied fruits and nuts make this fruitcake recipe from Tucsonans **Carl** and **Rose Gillin** a special treat. Directions for preparing it to age a year ahead are also given.*

2 cups cooking oil
2⅔ cups sugar
½ cup molasses
8 eggs
6 cups sifted flour
2 teaspoons baking powder
4 teaspoons each salt and cinnamon
2 teaspoons nutmeg
1 cup orange juice
1 cup brandy or sour mash whiskey
5½ cups seedless raisins
4 cups cut up dates
3 pounds mixed candied fruit
2 cups nuts broken in small pieces

Mix oil, sugar, molasses and eggs. Beat vigorously with spoon or electric mixer for 2 minutes. Sift together 4 cups flour, the baking powder, salt and spices, and stir in alternately with orange juice and whiskey. Mix 2 cups flour into the fruit and nuts. Pour batter over fruit, mixing thoroughly. Pour into 2 brown-paper lined tube pans. Place a pan of water on lower oven rack. Bake cakes at 275 degrees for 3 to 3½ hours.

After baking, let cakes stand overnight before taking from pans. Remove paper and store by wrapping tightly in aluminum foil. Place in covered fruitcake cans and store in cool place to ripen.

Preserving fruitcake

To preserve this fruitcake for the next year, wrap cooled baked cake in clear plastic wrap; overwrap with heavy-duty foil. Open the top and put cheesecloth or dish towel on top and pour over the cloth at least 2 shots of whiskey or brandy. Cover with outside wrapping. The cake should be stored in a tin container with a lid.

Every 6 weeks through November of the next year, repeat with more whiskey or brandy, except in June or July when the cake should be turned over. Turn back 6 weeks later and proceed with the liquid formula. Turn over again in November. Using the sour mash leaves a tart taste; the brandy a sweet taste.

Holiday fruit and nut cake

*For holiday serving, **Idola Lemen** of Tucson enjoys making this recipe from an Illinois friend.*

1 cup sugar
1 cup flour
3 teaspoons baking powder
¼ teaspoon salt
4 eggs, separated
2 teaspoons milk
1 pound pecans
1 pound seeded dates
1 teaspoon vanilla

Mix dry ingredients. Beat egg yolks with milk. Add nuts, dates and vanilla. Fold in egg whites, beaten stiff but not dry. Bake for 1 hour or until done in an oiled and floured loaf pan. For glass pan, bake at 325 degrees; for tin pan, bake at 350 degrees.

Date-nut coffeecake

Tucsonan **Betty Cowley** *was given this tasty recipe by a Michigan neighbor.*

3 cups flour
1 cup granulated sugar
1 teaspoon cinnamon
½ teaspoon salt
1 cup firmly packed brown sugar
½ cup lard
1 cup chopped dates
1 cup chopped nuts
1 teaspoon soda
1 cup sour milk

Sift together flour, granulated sugar, cinnamon and salt. Stir in brown sugar and cut in lard until mixture resembles cornmeal. Reserve ¼ cup of dry mixture for topping. To remaining ingredients add the dates, nuts and the soda dissolved in the sour milk. Turn into greased 9-by-13-inch pan. Sprinkle with reserved topping. Bake at 350 degrees for 35 to 40 minutes. It's best served warm.

Arizona date delight

An unusual date-nut cake that requires little mixing was worked out by Tucsonan **Dorothy E. Kilmer.**

1½ cups brown sugar
1½ cups warm water
1 cup white sugar
1 cup flour
1 teaspoon baking powder
¼ teaspoon salt
1 cup dates, cut up and packed
1 cup nuts, cut up
1 teaspoon vanilla
1 cup milk
Whipped cream

Put brown sugar and water in a 9-by-13-inch pan. Do not mix. In a bowl, mix white sugar, flour, baking powder, salt, dates, nuts, vanilla and milk. Pour on top of brown sugar and water. Do not mix. Bake at 350 degrees for 1 hour or until brown. Cut into squares. Flip upside down and top with whipped cream.

Westhaven cake

Joan Halper *of Tucson substitutes pecans for walnuts in this handsome cake from a New Jersey aunt.*

2 cups dates, cut up
1 cup water
1 cup sugar
½ cup shortening
2 eggs
1 teaspoon vanilla
1¾ cups all-purpose flour
1 teaspoon salt
1 teaspoon baking soda
1 teaspoon cocoa
¾ cup chocolate bits
½ cup chopped pecans
Confectioners' sugar

Soak dates in the water. Bring to a boil, drain and reserve the liquid. Cream the sugar and shortening. Add eggs and vanilla. Beat well. Sift flour 3 times with salt, baking soda and cocoa. Add to shortening mixture, alternately with the date liquid, beating until smooth. Stir in the chopped dates. Pour into a well-greased oblong baking pan and sprinkle chocolate bits and chopped nuts over batter. Bake at 350 degrees for 45 minutes. Cool and dust with confectioners' sugar. (If glass dish is used, reduce oven temperature to 325.)

Date soufflé

*Tucsonan **Virginia Whitlow** was given this recipe at a wedding shower years ago and still enjoys using it.*

1 tablespoon all-purpose flour
1 teaspoon baking powder
½ cup granulated sugar
¼ teaspoon salt
3 eggs, well beaten
1 cup dates, chopped
   mixed with a little flour
½ cup nuts, coarsely chopped

Mix flour, baking powder, sugar and salt. Mix in the eggs. Fold in chopped dates and nuts and bake in greased 8-by-8-inch pan at 350 degrees for 20 minutes or until firm.

Skillet date pudding

*This has been a favorite of Tucsonan **Carmen Lilleboe's** family for many years. It's nutritious, filling and easy to make.*

4 tablespoons butter
½ cup granulated sugar
1 egg
1 cup chopped dates
2 cups flour
2 teaspoons baking powder
½ teaspoon salt
¾ cup milk
2 cups brown sugar
¼ cup butter
2 cups water

Cream butter and granulated sugar; stir in egg. Stir in dates. Sift together the flour, baking powder and salt, and add dry ingredients to first mixture alternately with the milk. Place brown sugar, butter and water in skillet and heat to boiling. Drop date batter into syrup by spoonfuls. Turn heat down, cover and cook for 15 minutes. Serve plain or with cream. Serves 8 or more.

Date-nut pudding

*Tucsonan **Mary Boyer** brought this holiday dessert idea with her from Illinois.*

1 cup soda crackers, rolled fine
½ pound dates
1 pound nuts, chopped fine
2 cups sugar
2 teaspoons baking powder
1 pound nuts, chopped fine
6 eggs, separated

Mix crackers, dates, nuts and sugar sifted with baking powder. Add beaten egg yolks. Mix with heavy spoon or with hands. Add stiffly beaten egg whites. Line large loaf pan with waxed paper and lightly grease. Bake at 325 degrees for 35 to 40 minutes. Serve with whipped cream if desired.

Edna Snider's date pudding

*Tucsonan **Gina Skelley** has found this recipe can be baked while the rest of the meal is being eaten and the table cleared away. Serve it hot with unsweetened whipped cream.*

1 cup chopped dates
1 cup broken pecans
1 cup soft bread crumbs (packed)
¾ cup sugar
3 eggs, separated
   Few drops vanilla
   Whipped cream

Gently mix dates, nuts, bread crumbs, sugar, beaten yolks and vanilla. Fold in whites beaten to hold peaks. Pour into greased 8-by-8-inch glass casserole; bake at 275 degrees for 35 to 45 minutes. Should be light to medium brown on top, still soft inside. Serves 6 to 8. Good cold or hot with whipped cream.

Double date pudding

*Tucsonan **Ruth Schoonover's** recipe has dates in both pudding and topping.*

2 cups boiling water
2 cups chopped dates
2 teaspoons soda
2 tablespoons butter or margarine
2 beaten eggs
2 cups sugar
½ teaspoon salt
1 teaspoon baking powder
2½ cups flour
1 teaspoon vanilla
½ cup chopped pecans

Pour boiling water over dates. Cool slightly and add soda, butter and eggs. Sift sugar, salt, baking powder and flour. Add to date mixture and beat. Add vanilla and nuts. Bake at 350 degrees in 9-by-13-inch pan. Test after 40 minutes as may need longer baking. Top with date topping.

Topping: Cook ¾ cup boiling water, 1 cup dates and 1 cup sugar until of spreading consistency. Add 1 teaspoon vanilla, 1 cup pecans and 1 tablespoon butter. Cool and spread on cut cake with whipped cream and a maraschino cherry on each piece.

Date snack

*Tucsonan **Carrie Peskin's** favorite date snack (nosh) comes from her daughter Rosalind.*

½ cup sugar (scant)
1 cup cut-up dates
1 cup cut-up pecans
3 eggs

Put sugar, dates and pecans in a bowl and mix with hands, coating dates and nuts well with sugar. Beat the eggs well and mix into date mixture. Grease small muffin tins and spoon ⅞ full. Bake at 325 degrees for 25 minutes. Makes 24.

Hawaiian hermits

*Veteran cookie-maker **Geneva Melgren** of Tucson has long enjoyed making these cookies, especially for Christmas giving.*

2½ cups flour
1½ cups sugar
1 teaspoon soda
1 cup butter
½ teaspoon allspice
1 teaspoon cinnamon
8 ounces dates, cut up
1 cup nuts, chopped
3 eggs

Mix dry ingredients, working in butter with pastry blender, as for pastry, until very fine. Add dates, nuts and eggs and mix well. Roll in small balls or drop by teaspoonfuls on buttered cookie tin. Bake for 12 minutes at 350 degrees. Remove from pan to cool. Makes 6 dozen.

Quick date bars

***Susanne S. Cheung** of Phoenix likes to make these when she has a craving for sweets.*

1 cup buttermilk pancake mix
¼ cup plus 2 tablespoons brown sugar
1 egg
¼ cup shortening
1 teaspoon vanilla
¼ cup milk
⅓ cup chopped dates
⅓ cup chopped nuts

Place pancake mix, sugar, egg, shortening, vanilla and milk in medium-sized bowl. Beat until smooth, about 2 minutes. Stir in chopped dates and nuts. Bake at 350 degrees in greased 7-by-11-inch pan for 20 to 25 minutes. Cut while warm into squares. Serve plain or rolled in confectioners' sugar. Makes 1½ dozen bars.

Date bars

*Tucsonan **Selma Hoefle** sends another version of date bars, this one with lemon flavoring. It came from her sister in Illinois and is a ribbon winner.*

3 eggs, separated
2 tablespoons cold water
1 cup sugar
1 cup flour
1 teaspoon baking powder
½ teaspoon salt
1 cup chopped nuts
1 cup chopped dates
1 teaspoon lemon extract

Beat yolks, water and sugar until light-colored and thick. Fold in flour which has been sifted with baking powder and salt. Fold in beaten egg whites; add nuts and dates and lemon extract. Pour into 9-by-13-inch greased pan. Bake at 350 degrees for 25 to 30 minutes. Cool and cut in bars and roll in powdered sugar.

Date-coconut cookies

***Nan Jones** of Tucson shares this simple but delicious cookie recipe that came to her from Texas. The cookies have excellent keeping qualities.*

1 pound dates, cut up
1 cup sugar
2 eggs, beaten
3 or 4 teaspoons butter
1 teaspoon vanilla
1 cup chopped nuts
3 cups crisp rice cereal
Grated coconut

Cook dates, sugar, eggs, butter and vanilla for 7 minutes in a heavy skillet, stirring, until thick. Dates will be soft and mushy. Add nuts and cereal. Drop from teaspoonful on coconut and roll into balls. Makes 50 to 60 cookies.

Sour cream date dreams

Sour cream gives these date cookies an added fillip.

¼ cup shortening
¾ cup brown sugar
½ teaspoon vanilla
1 well-beaten egg
1¼ cups flour
¼ teaspoon baking powder
¼ teaspoon cinnamon
½ teaspoon baking soda
¼ teaspoon salt
⅛ teaspoon nutmeg
½ cup sour cream
⅔ cup chopped dates
Pecan halves

Thoroughly cream together the shortening, sugar and vanilla. Add egg; mix well. Sift dry ingredients together. Add to shortening mixture alternately with sour cream. Stir in dates. Drop from teaspoon onto greased cookie sheet. Top each cookie with a pecan half. Bake at 400 degrees for about 10 minutes. Makes 3 dozen.

Date-pecan delight

*When you can gather fresh dates from your own tree, this stuffed date recipe is at its best, **Nada Creason** of Mesa has found.*

Select nice-sized fresh dates. Pit and carefully fill each date with a pecan half. Press date together. Melt dipping chocolate over hot water and drop dates, one at a time, into chocolate. Remove with a small fork and place on waxed paper until cool.

Date sticks

An attractive combination of dates and nuts is this suggestion from Tucsonan **Lunette D. Cole.**

1 cup sugar
1 cup chopped dates
¾ cup flour
1 cup chopped nuts
2 eggs, beaten
Pinch soda
Confectioners' sugar

Mix all ingredients except powdered sugar and transfer to a greased and floured 8-inch square cake pan. Bake at 350 degrees for 20 minutes until lightly browned. Cool slightly and cut in pan in stick-shape in desired size. Roll each stick in powdered sugar.

Date-filled drop cookies

Carline Marsh *of Morenci has found it wouldn't be Christmas in her family without these festive cookies.*

1½ cups shortening
3 cups brown sugar
3 eggs
¾ cup water
1½ teaspoons vanilla
5¼ cups flour
1½ teaspoons baking soda
1½ teaspoons salt
⅛ teaspoon cinnamon

Mix well the shortening, sugar and eggs. Stir in water and vanilla. Sift together and stir in flour, baking soda, salt and cinnamon. Chill for 1 hour. Drop by teaspoonfuls on ungreased baking sheet. Place ½ teaspoon date filling on dough and cover with ½ teaspoon dough. Bake at 400 degrees for 10 to 12 minutes.

Date filling: Cook until thick, stirring constantly, 2 cups dates, cut small; ¾ cup sugar, ¾ cup water and ½ cup chopped nuts. Cool. Makes about 9 dozen.

Note: Recipe can easily be cut by ⅓, to make 6 dozen: 1 cup shortening, 2 cups brown sugar, 2 eggs, ½ cup water, 1 teaspoon vanilla, 3½ cups flour, 1 teaspoon soda, 1 teaspoon salt and good pinch of cinnamon.

Party date bars

Betty M. Wiltfang *likes to keep these great cookies on hand because she never knows when company may come to visit her in Tucson from her former South Dakota home.*

Filling:
2 cups pitted dates
½ cup sugar
Grated peel of 1 orange
¼ cup orange juice
½ cup chopped nuts

Crumb base:
1 cup sifted flour
½ teaspoon soda
½ teaspoon salt
1 cup oatmeal, quick or old-fashioned (uncooked)
1 cup firmly packed brown sugar
½ cup butter or margarine, melted

For filling, combine dates, sugar, orange peel and orange juice in medium saucepan. Cook over low heat, stirring occasionally until thickened. Cool. Stir in nuts.

For crumb base, sift together the flour, soda and salt. Add oats and sugar and mix well. Stir in the butter and mix until crumbly. Firmly press ⅔ of mixture into greased 8-inch square baking pan. Spread filling evenly over crumb base. Top with remaining crumb mixture; pack lightly. Bake at 350 degrees for 30 minutes. Cool and cut in bars. Makes about 2 dozen.

Date-filled cookies

Jacquie Kata of Tucson finds these old family favorites always good eating.

½ pound chopped dates
¼ cup water
½ cup sugar
½ cup chopped pecans
1 teaspoon lemon juice
8 ounces cream cheese
½ pound butter or margarine
2½ cups sifted flour

Prepare filling by cooking dates, water and sugar over low heat for 2 to 3 minutes. Add chopped pecans and lemon juice.

Cream softened cream cheese and butter together and work in flour for a dough. Roll out dough between pieces of waxed paper. Cut into circles with cutter or glass (1½ to 2 inches in diamater). Place half the circles on greased cookie sheet. Spoon 1 scant teaspoon filling on each center. Place another circle of dough on top and press to seal with fingers; then crimp edges with tines of fork. Bake at 350 degrees for about 12 minutes.

Stuffed date drops

*A Christmas treat in Tucsonan **Loydeen Waitt's** family is stuffed date cookies.*

1 pound pitted dates
1 nut half for each
¼ cup shortening
¾ cup brown sugar
1 egg
1¼ cups sifted all-purpose flour
1 teaspoon soda
1 teaspoon baking powder
¼ teaspoon salt
½ cup thick sour cream

Stuff each date with a nut. Cream shortening and sugar. Beat in egg. Sift dry ingredients together. Blend into egg mixture alternately with sour cream. Add dates. Stir until dates are coated with batter. Drop on greased cookie sheet, 1 stuffed date for each cookie. Bake at 375 degrees for 10 to 13 minutes. Cool. Frost with browned butter icing. Makes 4 dozen.

Browned butter icing: Lightly brown ½ cup butter in skillet. Beat in 3 cups powdered sugar and 1 teaspoon vanilla. If too thick, thin with water.

Dates for home gardeners

Pollination. Bees are not dependable pollinators because date blossoms have no scent to attract the insects, and female date flowers must be pollinated by hand. Generally this is done in late March or April. If there is no male tree nearby, nurserymen can advise those interested on where to get the necessary male flowers.

Thinning. Usually in June, when the fruit is about ¼ inch in diameter, a portion of the fruit should be removed to improve the size and quality of the fruit, to lighten the bunch and to ensure a good crop the next year. Generally the smaller bunches are removed, and then the remaining bunches are thinned by removing the center of the bunch and/or cutting the tips back, or even by removing individual fruits on the strands.

Supporting. When the fruit becomes heavy enough to force the bunches down through the leaves, the bunches may be tied to adjacent leaves for support.

Reducing spoilage. Rain can cause injury to green fruit by producing small cracks that heal with scar tissue, and to ripe fruit by producing larger cracks that do not heal and allow fungi, yeasts and acid-forming bacteria to enter the fruit, causing fermentation, souring and decay. To reduce injury during the rainy season, each bunch of dates can be covered with a commercial waterproof paper protector that contains a small amount of wax. The paper is shaped into a tube and tied around the bunch, allowing the paper to extend almost to the bottom of the dates.

Paper covers also protect the fruit from damage by birds and reduce injuries from wasps and bees, but the covers should be lifted from time to time to allow air to circulate.

Harvesting. When fruit starts to ripen, it becomes translucent and the flesh becomes soft

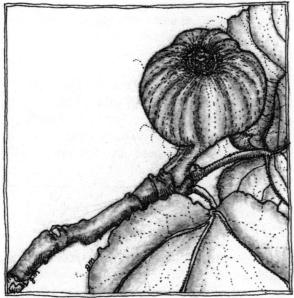
Fig

and pliable, turning a tannish or reddish color, depending on variety. The fruit should be picked as it ripens. Sour, overripe and diseased fruit should be picked from the clusters and any fallen fruit should be removed from the ground beneath the palms to control insects.

Cleaning. Dates should not be washed. Spread them on a clean, moist terry cloth towel on a tray and shake or roll dates to remove dust and dirt.

Since dates often contain outside contamination, they can be given a heat treatment after cleaning. Tucsonan Harvey Tate recommends this simple method: After cleaning, spread dates on cookie sheets. At dinnertime, preheat the oven to 250 degrees, turn it off and set the dates in the oven, leaving them there overnight as the oven cools. The next morning, remove dates from the oven, preheat again to 250 degrees, turn off heat, return dates to oven and leave while the oven cools. Dates are now ready to refrigerate until ready to eat out of hand or use for baking.

Preserving. For storage in a cool, dark place, dates need processing for longer periods than the method given above. A dehydrator or sun heater may be used. Or the fruit may be canned in jars in a pressure cooker for 15 minutes at 10 pounds pressure.

Figs

Botanical name: *Ficus carica*

Readers of the Bible are well-versed in stories about the ancient fig. Among them: The aprons that Adam and Eve sewed together were made of fig leaves; Canaan's favorite food products included figs; Saint Augustine sat under a fig tree when he contemplated the Scriptures.

In Eastern mythology, the fig tree also is associated with religion. Siddhartha Gautama, founder of Buddhism, sat under a fig tree as he meditated.

In modern times, though admired, the fig has fallen from its once prominent place on the table. It is now produced in relatively small amounts, with most of the commercial crop coming from California.

Figs are a good source of B vitamins, especially folic acid, potassium and magnesium. Dried ones, however, have a concentrated sugar content, with a calorie count close to 300 per 100 grams. Dieters, beware.

The fig is unique. Unlike other fruits, its blossoms are not displayed for all to see, but are enclosed inside the fleshy center of the fruit.

Best for growing in the southern half of the state, especially in lower elevations, are the purple-black Black Mission and the white Kadota and Canadria. The fruit ripens in late spring and again in summer. Canadria is a white fig that produces well under minimum chilling conditions. Brown Turkey prefers 2,000 to 3,000 feet. Texas Ever-Bearing also does well in the Tucson area.

Fig-stuffed pork loin roast

A good fig crop can lead to a supply of dried figs to enjoy for many months. Here's an unusual idea for them:

½ cup chopped celery
¼ cup chopped onion
 1 clove garlic, chopped
 2 tablespoons butter or margarine
½ cup soft bread crumbs
¼ cup chopped walnuts
¾ cup diced, dried figs
 1 teaspoon salt
½ teaspoon ground sage
¼ teaspoon pepper
 1 5-pound loin of pork, with deep pockets
   cut between chops (about 10)
 3 to 4 tablespoons honey (about)

Sauté vegetables in butter until almost tender; combine with crumbs, walnuts, figs and seasonings. Pack stuffing into pockets between chops. Place on rack in shallow roasting pan. Roast at 325 degrees for 35 minutes per pound, or until meat thermometer registers 170. (If stuffing starts to brown too much, cover lightly with foil.) A half hour before end of roasting time, spoon some honey over roast. Baste once with gravy made from roast drippings, if desired. Serves about 8.

Stewed figs

*For **Marian Strang** of Tucson, a tree full of fresh figs will surely mean it's time to stew them. This one of her husband's favorite fig dishes.*

To stew, wash and cut up figs, and put them in a saucepan with a little water, a slice or two of lemon or a stick of cinnamon. Stir in a little sugar. Cook and stir over low heat until figs are tender.

Freezing figs

Select tree-ripened, soft-ripe fruit, making sure the figs have not become sour in the center. Sort, wash and cut off stems. Peel if desired. Slice or leave whole.

Syrup pack. Use 35 percent syrup (dissolve 2½ cups sugar in 4 cups hot water), plus ¾ teaspoon crystalline ascorbic acid or ½ cup lemon juice. Pack figs into containers and cover with cooled syrup, leaving head space. Seal and freeze.

Unsweetened pack. Pack into containers, leaving head space. Cover with water or not as desired. For light-colored figs, if water is used, add ¾ teaspoon crystalline ascorbic acid to each quart of water. Seal and freeze.

Crushed. Prepare figs as directed and crush coarsely. Mix ⅔ cup sugar and ¼ teaspoon crystalline ascorbic acid, and combine with 1 quart prepared fruit. Pack into containers, leaving head space. Seal and freeze.

Fig jelly roll

If you've made plenty of fresh fig jam (see page 105), you may wish to turn some of it into desserts like this suggestion from **Lorene Kececioglu** *of Tucson.*

3 large eggs (⅔ cup)
1 cup sugar
5 tablespoons water
1 teaspoon vanilla
1 cup sifted flour
1 teaspoon baking powder
¼ teaspoon salt
    Fig jam

Beat eggs until thick. Gradually beat in sugar. Then beat in water and vanilla. Sift flour, baking powder and salt, and beat into egg mixture all at once, just until smooth. Pour into greased and floured, 15½-by-10½-inch jelly roll pan. Bake until cake tests done (375 degrees for 12 to 15 minutes). Loosen edges and immediately turn upside down on towels sprinkled with confectioners' sugar. Spread cake at once with fig jam and roll up, beginning at short end. Wrap in the towel and cool ½ hour. Slice to serve.

Fig pie

Irene Searcy *of San Manuel recommends this pie for anyone with extra figs.*

1 quart ripe figs
1 6-ounce can frozen lemonade concentrate
2½ tablespoons quick-cooking tapioca
1 tablespoon butter or margarine
    Unbaked pastry for 1 large or 2 small pies

Wash, remove ends from figs, mash or slice. Mix with concentrate and tapioca and pour into unbaked pie shells. Dot with butter. Add top crust, slash and bake at 350 degrees for about 45 minutes or until lightly browned.

Fig cake filling

Tina Poindexter *of Safford uses this cake filling for white cake.*

Combine 1 cup ground pecans, 1 cup figs (or dates), chopped, and 2 egg whites, beaten stiff. Add enough honey to allow filling to spread nicely.

Fig-rhubarb pie

Of special delight to both fig and rhubarb lovers is this pie from **Mary A. Hardy** *of Tucson.*

1 9-inch baked pie shell
2 tablespoons butter
2 cups sliced fresh figs (unpeeled)
2 cups finely sliced rhubarb (fresh or frozen)
1¼ cups sugar
3 tablespoons cornstarch
¼ cup cream, evaporated milk or half-and-half
2 large eggs, separated
1 teaspoon vanilla
4 tablespoons sugar

Melt butter in saucepan, add figs, rhubarb and 1¼ cups sugar. Blend thoroughly; cook and stir over medium heat until sugar melts and figs and rhubarb are soft. Combine in bowl the cornstarch, beaten egg yolks and cream, and add egg mixture to fig mixture. Cook and stir until thickened. Stir in vanilla. Pour into baked pie shell.

To make meringue, beat egg whites until stiff, gradually beating in 4 tablespoons sugar. Top filled pie shell with meringue. Bake at 300 degrees for 15 minutes, or until lightly browned. Makes 1 9-inch pie.

Fig bars

Dried figs go into these traditional fig bars, enhanced with grated lemon peel.

Cookie:
¾ cup shortening
¾ cup sugar
1 egg
½ teaspoon vanilla
½ teaspoon grated orange peel
2 cups sifted flour
1½ teaspoons baking powder
¼ teaspoon salt
4 teaspoons milk

Filling:
1 cup dried chopped figs
Grated peel of 1 lemon
1 cup boiling water
½ cup sugar
2 tablespoons flour

Cream shortening and sugar. Add egg and beat until mixture is light and fluffy. Add vanilla and grated peel and mix thoroughly. Sift dry ingredients and mix into creamed mixture with the milk. Divide dough in half. Chill 1 hour.

Prepare filling by adding figs and grated peel to boiling water. Mix sugar and flour and add some of hot mixture. Stir flour mixture into remaining hot mixture. Cook and stir for 5 minutes and cool.

Work with ½ dough at a time, keeping the rest chilled until needed. Roll dough ⅛-inch thick and cut in long strips 3 inches wide. Spread half the dough strips with the filling and cover with remaining strips. Press edges together with a fork. Cut strips in 2-inch lengths. Bake on greased cookie sheet at 350 degrees for 20 minutes. Makes 2 dozen bars.

Fig pinwheels

Anyone with a fig tree will love using this recipe from **Carole Waina** *of Tucson.*

1 cup figs, put through food chopper
1 cup nuts, chopped
¼ cup water
1 cup granulated sugar
½ cup butter or margarine
½ cup brown sugar
1 egg
2 cups flour
½ teaspoon soda
¼ teaspoon salt

Combine chopped figs, water and ½ cup sugar. Cook until thick. Cool. Fold in nuts. Cream butter, remaining granulated sugar and brown sugar. Mix in egg. Sift flour, soda and salt and add to first mixture. Roll on floured board or cloth. Spread with fig mixture; roll up like jelly roll. Chill. Cut in slices and bake at 400 degrees for 12 minutes.

Fig-orange conserve

Harriet Smith *of Green Valley concocted this conserve to use some of the crop from her huge fig tree.*

4 cups fresh figs (about a dozen),
  peeled if desired
1 small slice of lemon, finely chopped
2 thin slices orange, finely chopped
¼ cup sugar
¼ cup chopped nuts
2 ounces crystallized ginger, diced

Combine figs, lemon, orange and sugar and let stand 3 to 4 hours. Cook and stir in saucepan over medium heat until partially cooked. Add nuts and diced ginger and continue cooking for a few more minutes. Pour in sterilized jars and add paraffin. Recipe can be doubled or tripled.

Mock strawberry preserves

Whenever the fresh figs are in, you can be sure someone will be needing this popular recipe. The Star prints it almost as often as it does recipes for prickly pear jelly!

6 cups fresh ripe, peeled figs
3 cups sugar
1 family-size or 2 regular-size packages
   strawberry gelatin

Cook ingredients in a large pot, very slowly, stirring constantly. Cook until thickened, about 40 minutes. Pour into hot pint jars and seal. Makes 3 pints. Preserves taste like strawberries.

Fig chutney

*Once, when her figs were coming in by the dishpanful, Tucsonan **Marjorie Hefty** obtained several good recipes from friends. Among them, this chutney.*

1 medium onion, diced
1 small clove garlic, smashed
1 cup seedless raisins
8 cups diced figs
2 tablespoons chili powder
2 mangos, peeled and diced (optional)
1 cup crystallized ginger, chopped
2 tablespoons mustard seed
1 tablespoon salt
1 quart vinegar
2¼ cups brown sugar

Put onion, garlic and raisins in large kettle. Mix diced figs with remaining ingredients and add to onion mixture. Mix well. Simmer, stirring, an hour until deep brown and rather thick. Remove garlic clove. Pack into sterilized jars and seal at once.

Fig jam

Eunice I. Johnson of Phoenix shares this jam recipe that gets an extra fillip from a touch of apricot brandy.

Use Brown Turkey figs, some very ripe and some half-ripe. Peel them and cut them up. Measure and put in a large kettle. Add ¾ cup sugar to each cup of figs. Let mixture stand for 15 minutes, and stir them well.

Cook on a low fire, stirring frequently, for 25 minutes. Turn off fire and let mixture set 2 to 3 minutes. For each jar, stir in about 1 tablespoon apricot brandy or a good dash of cinnamon. Cover jars with paraffin.

Spiced pickled figs

*This recipe was given to **Juanita Stone** of Scottsdale to help her use the figs from her tree. It's worth the time it takes to prepare it.*

10 pounds firm, ripe figs
   Whole cloves
7 pounds sugar
1 quart cider vinegar
1 teaspoon whole allspice
4 2-inch pieces cinnamon

Wash figs; leave whole and unpeeled. Stick 2 cloves in each fig. Set aside to use later. Combine sugar, vinegar, allspice (tied in a bag) and cinnamon in a large kettle (enamel or stainless steel). Mix well and bring to boil. Add figs and cook for 5 minutes, uncovered. Remove from heat and cool. Cover and let stand overnight.

Remove figs from syrup the next day. Bring syrup to boil, pour over figs and let stand for 48 hours. Remove figs from syrup again. Bring syrup to boiling point, add figs and boil about 30 seconds or until figs are hot. Ladle into hot, sterilized 1-pint jars. Seal. Let stand at least 4 weeks before using. Makes 8 to 10 pints.

Spiced figs

Leslie Daniels of Tucson uses this smaller-scale spiced-fig recipe with great success.

3 quarts washed fresh figs, unpeeled
4 to 6 cups sugar
1 cup water
1 cup white vinegar
1 or 2 sticks cinnamon
Several whole cloves

Combine sugar, water and vinegar in large saucepan and boil until sugar is dissolved. Place spices in tea ball or tie in cheesecloth and add to syrup. Add figs and boil for 10 minutes. Cover and set aside. The next day, bring to a boil and cook for 10 minutes. Cover and set aside. On the third day, bring to a boil, cook for 10 minutes and seal in jars or store in refrigerator. Makes 10 to 12 ½-pints.

Fig daiquiris

Jane R. Stinson of Tucson shares this suggestion for a drink made with figs, fresh or frozen. The drink is a pretty shade of pink.

Pick ripe figs (the Stinson figs are black figs), put in plastic bag and sprinkle with 2 teaspoons sugar. Store in refrigerator or freezer. To make drinks, fill blender ½ to ⅔ full of ice, add 10 figs and 6 packets of daiquiri mix, along with 8 1½-ounce jiggers of rum. Makes 12 daiquiris.

Jujubes

Botanical name: *Ziziphus jujuba*

Known as the Chinese date, the jujube (pronounced "jew-jewb," with the accent on the first syllable) is a slow-growing tree that produces reddish brown, single-seeded, edible fruits from olive-size to 1½ inches in diameter. The trees can grow 30 feet tall and tolerate heat and cold.

Blossoms appear in late spring, but are not much to speak of. They are small and yellowish. The leaves turn golden before they drop off in the fall. The trees are more often grown for decorative purposes than for their late summer fruit, although the fruit has a pleasant flavor resembling apples. Nutritionally, the fruit also resembles apples. Jujubes can be eaten fresh, made into jam or used in other ways. They are an especially good fruit for drying and can be used in the same way as prunes.

Jujube butter

Like apples, jujubes make excellent spiced butter.

3 cups sieved jujube pulp
Sugar
½ teaspoon cinnamon
¼ teaspoon nutmeg
⅛ teaspoon ground cloves
Juice of 1 lemon

Cover fruit with water and simmer. When tender, rub through sieve or colander to remove skin and seeds. Add sugar to taste (start with 1 cup). Stir in spices and lemon juice and simmer until thick. Makes about 2 cups butter.

Stewing jujubes

*Tucsonan **Esther Tang,** who enjoys the fruit Chinese-style in clear soups, also likes to stew dried or fresh jujubes.*

Dry by spreading fresh fruit on trays and placing in the sun, a dehydrator or in the oven.

Eat as dried fruit, or simmer in water with a slice of lemon. Add sugar or honey to taste. Use as you would prunes.

Stew fresh jujubes in water with lemon, and add sugar or honey to taste. Use as you would other stewed fresh fruit.

Jujube mince meat

If you have jujubes in your yard, gather some when ripe (usually in September) and try them in one of these suggestions.

1 pint green tomatoes
1½ pints jujubes, seeded
1½ cups sugar
½ cup vinegar
1 teaspoon each ground cinnamon, nutmeg and cloves
1 teaspooon flour
1 cup raisins

Grind tomatoes and jujubes fine and combine with sugar, vinegar and spices. Simmer for 30 minutes. Stir flour into a small amount of water and stir into the fruit mixture along with the raisins. Simmer for 15 minutes. Use for making pie.

Loquats

Botanical name: *Eriobotrya japonica*

Also called Japanese plum, the loquat tree is medium in size and has large, leathery leaves 6 or more inches long. It grows in Japan, China and the Mediterranean region, as well.

The tree, which prefers low to moderate elevation, ordinarily is used as an ornamental. But in late spring, yellow to orange, inch-long fruit develops that is delicious. It has several small seeds in the center, which are removed before using. The variety recommended for desert growing is Champagne.

Loquats can be eaten fresh, but are usually cooked or made into preserves. Loquats are a good source of potassium and contain a fair amount of vitamin A. They are low in vitamin C.

Loquat jelly

Hanna Lundberg *of Tucson suggests this method for making jelly with loquats.*

Pick loquats when partially turned in color, but still hard. Wash and remove blossom ends. Cover with cold water. Cook slowly until pulp is very soft. Drain in a jelly bag. Cook juice until thick and pink in color. Measure 1 cup juice to 1 cup sugar and boil rapidly for about 15 minutes. Pour into sterilized jars.

Loquat preserves

When their loquat tree bears fruit, as it does some years, **Nicholas and Charlotte Furlong** *of Tucson make preserves.*

3½ cups prepared loquat pulp
⅓ cup lemon juice
6½ cups sugar
½ bottle liquid pectin

To prepare fruit, wash and steam briefly. Remove peeling with a paring knife, cut in half and take out the seeds. Put fruit in open kettle with lemon juice and sugar and bring to rolling boil, stirring constantly. Boil 1 minute, take off heat and add liquid pectin. Skim off foam. Stir for 5 minutes as fruit cools. Pour in sterilized jars and add paraffin.

Upside-down loquat cake

The fruit is not peeled in this recipe from **Leo Della Betta** *of Tucson.*

1 box yellow or spice cake mix
4 tablespoons butter
4 tablespoons brown sugar
    Juice of ½ a small lemon
    Loquats, halved and seeded

Prepare cake mix according to directions. Mix butter and brown sugar in a utility pan (9-by-13 inches), add lemon juice and cook and stir over low heat until mixed and syrupy.

Cover the syrup in the bottom of the pan with halved and seeded loquats, cut side down. (No need to peel them). Pour cake batter over the fruit and bake as directed on package. Remove from oven when done and let stand about 15 minutes. Then invert carefully (there will be some juice) over a large platter.

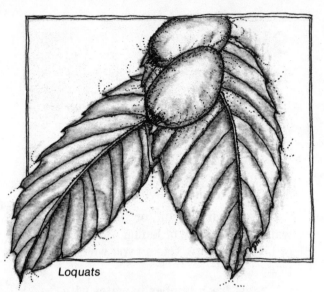

Loquats

Loquat jam

Another recipe from Tucsonan **Leo Della Betta.**

4 cups prepared loquats
1½ cups sugar
¾ cup water
½ cup lemon juice

Wash, seed and chop enough loquats to make 4 cups. There is no need to peel the loquats. Put the loquats through a meat grinder (this is a bit messy) or chop them a few at a time on a large chopping board. (Save and include the juice in the 4 cups.)

Mix the pulp and juice with the sugar, water and lemon juice in a large pan. Cook until thickened (225 degrees on a candy thermometer). Pour into jars and seal.

Note: The proportion of sugar is smaller than that in most jam recipes, but increasing it will cause sugar crystals to develop in the jam within a few months. For most people, the jam is sweet enough with less sugar.

Stewed loquats

To stew loquats, peel and remove the large dark seeds. Cook in small amount of water until tender, and sugar to taste.

Melons

"Friends are like melons, shall I tell you why?
To find a good one, you must a hundred try."

So goes an old saying describing the difficulties of selecting a good melon — unless you cut it.

But once found, a slice of any kind of ripe melon is delicious to eat, and cooling to look at. Being about 90 percent water, it is also cooling in fact.

Melons belong to either the muskmelon group (such as cantaloupes) or the watermelon group. Both groups are members of the large gourd family *Cucurbitaceae* and are kin to the pumpkin, cucumber and squash.

Muskmelons probably originated in Persia. The "musk" portion of their name derives from their aroma that was considered similar to musk, the Persian perfume oil.

As early as 2400 B.C. Egyptians admired muskmelons and painted them on the walls of tombs. Later, Greeks, Italians and other Europeans enjoyed the fruit. Columbus brought the seeds to the New World, where they immediately became a delight of the Indians.

The name cantaloupe came later. It is derived from *Cantalupo*, an Italian castle where an apple-sized version of the fruit was grown in the 17th century.

In the same family with the cantaloupes are the late or "winter" melons, the casaba, honeydew, etc. They ripen in the late summer and keep better than cantaloupes. They lack the musky odor.

The watermelon's roots are in Africa, where the plants were grown at least 4,000 years ago. It was David Livingstone whose African explorations disproved a long-held view that watermelons originated in Asia. As did the muskmelon, the watermelon came to the Americas by way of Europe and the explorers.

Cantaloupes

Botanical name: *Cucumus melo cantalupensis*

Though most people serve cantaloupes chilled, their apricot-colored flesh is more flavorful at room temperature. They are generally eaten raw, in combination with other fruits or as a dessert with ice cream. They make an expensive but justifiably famous appetizer when combined with Italian prosciutto ham.

To grow cantaloupe successfully in the backyard garden is to ensure flavor beyond most available at the supermarket. The home gardener can wait to pick the cantaloupes until they are fully ripe — when the stem has cracked away from the melon.

Cantaloupes are not to be confused with Persian melons. While both have a netting on the skin, cantaloupes are round; Persian melons are larger and generally are globe-shaped.

Cantaloupes are a low-calorie delight high in vitamin A (more than 3,000 international units in a 100-gram or 3½-ounce serving). They also contain a good amount of vitamin C and potassium.

Cantaloupes grown in Yuma and Parker are excellent, but very good ones also are produced in the Salt River Valley, parts of Pinal County and in the Elfrida area. A popular variety for home growing is Top Mark.

Late melons

Botanical name: *Cucumus melo var*

Late melons, such as casaba, honeydew and Persian melons, can be grown. Some of their major characteristics:

• The round casaba has golden, light green or dark green rind with deep wrinkles. The flesh is golden to light yellow, depending on variety.

• The Crenshaw (sometimes spelled Cranshaw) is round at the base and slightly pointed at the stem end. The rind is gold and green, without netting. The flesh is a handsome salmon color.

• The honeydew has a smooth, creamy white skin, with flesh pale green to white. When ripe, it smells "as sweet as honey." The honey ball resembles the honeydew, but is smaller.

• The Persian melon is usually globe-shaped, with a fine netting over its dark green rind. The flesh is orange.

Like cantaloupes, Persian melons break cleanly from the stem when ripe. The other late melons may be tested by smelling the blossom end, which has a pleasant, fruity flavor when ready to eat.

Nutritionally, late melons offer some vitamin C and are low in calories. Those with salmon or orange-colored flesh are a good source of vitamin A. All are fair sources of potassium.

Watermelons

Botanical name: *Citrullis vulgaris*

Watermelons are a summertime fruit that can grow as far north as Massachusetts. In home gardens in the Sonoran Desert and nearby, they do quite well. Among varieties that have proved popular are the Sugar Baby and Yellow Baby (smaller watermelons), and Charleston Gray (20 pounds and over). A new one that is doing well is Jubilee.

When ripe, the underside of watermelon turns yellowish. This coloration is considered a more dependable method to determine when a watermelon is ready than thumping. Besides eating delight, a good-sized slice of watermelon contributes a fair amount of vitamin A. The vitamin C content, however, is only a fourth as high as that of other melons.

Melon chicken salad

A favorite summer salad with **Kathy H. Wilson** *of Chandler is this one with cantaloupe and chicken.*

1 cantaloupe, cut in cubes
Chopped cooked chicken in an amount
  equal to the cantaloupe (about 2 cups)
1 slice cheese, cut up (optional)
½ cup chopped celery
Pitted olives and/or chopped nuts
  (optional)
½ cup mayonnaise or salad dressing
1 tablespoon lemon juice

Combine cantaloupe, chicken, cheese, celery and pitted olives. Mix mayonnaise and lemon juice and then combine with fruit-chicken mixture. Serve on salad greens, for a main-dish salad. Serves 4.

Chantilly sauce

A delight to serve over melon and other fruits for a dessert salad is this fruit sauce flavored with orange liqueur.

Combine 1 8-ounce package cream cheese, softened, and 2 tablespoons sugar, mixing until well blended. Add 3 tablespoons orange liqueur and 2 tablespoons milk and mix at high speed in electric mixer until creamy. Serve over fruit. Can be made ahead and mixed until creamy just before serving. Makes 1¼ cups.

Cantaloupe and peach pie

Genevieve Sitarz *of Tucson finds fresh peaches combine well with cantaloupe in this two-fruit pie.*

Pastry for two-crust pie
2 cups thinly sliced cantaloupe
2 cups sliced peaches
1 cup sugar or ¾ cup mild honey
¼ cup unbleached flour
½ teaspoon salt (optional)
¼ cup sliced almonds
1 tablespoon butter
1 tablespoon sour cream (optional)

Place sliced cantaloupe and peaches in large mixing bowl and toss lightly with a mixture of sugar, flour and salt. Arrange fruit in pastry-lined pie pan. Sprinkle with nuts and dot with butter. Cover with top pastry and slash. Brush top with 1 tablespoon sour cream, if desired. Bake at 350 degrees about 40 minutes. Serve hot or cold.

Watermelon ice with fruit

For an unusual summertime cooling dessert, make watermelon ice and serve with other fruits.

3 cups watermelon purée (about 4 cups watermelon pulp)
¾ cup sugar
3 tablespoons lemon juice
⅛ teaspoon salt
1 teaspoon unflavored gelatin
¼ cup cold water
Assorted fresh fruits to serve 8 to 10

Blend watermelon pulp to make 3 cups purée. Stir in sugar, lemon juice and salt. In small saucepan, soften gelatin in cold water. Heat and stir until gelatin dissolves. Add gelatin mixture to melon mixture; blend thoroughly. Pour mixture into 9-by-9-by-2-inch pan; cover and freeze until partly frozen. Turn mixture into large chilled mixer bowl. Beat at medium speed of electric mixer until light and fluffy. Return to pan. Freeze until firm. Let stand 5 minutes before scooping. Makes 4 cups watermelon ice. Serve on dessert plates over individual servings of mixed fruits: pineapple cube, seedless grapes, sliced and pitted plumbs, etc. Serves 8 to 10.

Cantaloupes

Melon dessert salad

Tucsonan **Hazel Coatsworth** *suggests this flavorful combination of Arizona-grown fruits for a low-calorie salad or dessert.*

2 cups diced cantaloupe
2 pears, cored and diced
1 cup seedless grapes
2 tablespoons lemon juice
Salad greens, if desired

Combine ingredients and serve on salad greens. Serves 4.

Cantaloupe preserves

Enjoy cantaloupe preserves after the fresh season is gone, as suggested in this recipe from **Julia Pearson** *of Oracle.*

3 medium-size ripe cantaloupes
⅓ cup orange juice
1 orange and peeling of ½ orange
4 tablespoons lemon juice
1 1¾-ounce package powdered pectin
5½ cups sugar

Seed and peel melons, dice and put in 5-quart kettle. You will need about 5½ cups melon. Add orange juice and cook over low heat, stirring so it doesn't stick, for about 10 minutes. When melon is soft, mash with potato masher until broken up in fine pieces.

Peel orange and grind or cut into very small pieces. Grind in blender or chop fine half the orange peel. Add the lemon juice. Combine the orange mixture with the cantaloupe. Stir in pectin and bring to a boil. Add the sugar and bring to a rolling boil that cannot be stirred down. Cook over medium heat and stir occasionally, to keep from sticking, for about 25 minutes. Skim and ladle into sterilized pint or ½-pint jars. Makes about 4 pints.

Cantaloupe mold

As cooling as can be, this dessert mold recipe can be cut in half, if desired.

2 packages orange-flavor gelatin
1 cup boiling water
2 cups puréed cantaloupe (1 medium)
⅔ cups undiluted evaporated milk
2 tablespoons lemon juice
Melon balls, fresh mint

Dissolve gelatin in boiling water. Stir in cantaloupe. Chill until mixture is consistency of unbeaten egg white. Beat with rotary beater until smooth. Pour evaporated milk into ice-cube tray or 8-inch metal cake pan. Chill in freezer until soft ice crystals form around edges (10 to 15 minutes). Whip in small mixer bowl on high speed until stiff, about 1 minute. Add lemon juice. Whip very stiff, about 2 minutes longer. Fold into cantaloupe mixture. Spoon into 6-cup mold. Chill until firm, 2 to 3 hours. Serve garnished with melon balls and mint.

Nectarines

Botanical name: *Prunus persica nectarina*

For a long time, people thought a nectarine was just a fuzzless peach. There's even a widely held view that a nectarine is a cross between a peach and a plum.

Some authorities say neither is true. Like cherries, apricots and peaches, the nectarine is a member of the rose family. But just as a rose is a rose is a rose, a nectarine is a nectarine. Luther Burbank was one who believed that nectarines predated peaches.

The fruit has been a favorite of the Chinese and Egyptians for several thousand years. In the last decade or so, it has become popular in this country. California produces most of the commercial crop of this exotic, smooth-skinned fruit. With proper care, it can do quite well in home orchards of both the Sonoran Desert and higher portions of Southeastern Arizona.

In Maricopa County, Armking is a recommended variety. In the middle and high deserts, nectarines require warm microclimates. Dwarf varieties are available for home orchards.

Like peaches, nectarines can make summer salads, meat dishes, ice cream and pies more appealing. Among their nutrients, chalk up high credits for potassium and vitamin A. They can be canned, frozen or dried.

Nectarine salad

Linda Abrams of Tucson combines 4 cups cut-up fruits (nectarines, melons, bananas, oranges, apples, etc.) with 1 cup plain or berry-flavored yogurt, ½ cup raisins and ½ cup chopped pecans. Chill and serve for an interesting and pleasing fruit salad. Makes 6 cups.

Nectarine cake

When nectarines are ripe, it's time to put together this great fresh fruit-and-cake combo.

2 tablespoons melted butter or margarine
3 tablespoons sugar
½ teaspoon grated lemon peel
1 tablespoon lemon juice
½ teaspoon cinnamon
2 cups sliced nectarines, unpeeled
1 loaf-size package white or yellow
   cake mix
   Whipped cream

Melt butter in bottom of 9-inch round glass baking dish. Sprinkle with sugar, lemon rind, lemon juice and cinnamon. Arrange nectarine slices on top of sugar mixture. Prepare cake mix, pour over fruit. Bake at 350 degrees for 35 to 40 minutes. Cool on wire rack for 5 minutes, then run knife around edge and invert on serving plate. Serve with whipped cream. Makes 1 cake.

Nectarine pie

Tucsonan **JoAnne Jones** *finds this fresh-fruit pie suits her family just fine.*

Filling:
¾ to 1 cup sugar
2 tablespoons flour
1 teaspoon cinnamon
¼ teaspoon nutmeg
   Dash salt
6 to 8 nectarines, sliced thin
2 tablespoons butter

Pastry:
2 cups flour
2 teaspoons sugar
1¼ teaspoons salt
⅔ cup salad oil
3 tablespoons milk

To make filling, combine the sugar, flour, cinnamon, nutmeg and salt and mix with fruit.

To make crust, sift into a 9-inch pie plate the flour, sugar and salt. Whip the oil with the milk and pour over flour mixture. Mix with fork until dampened. Press ⅔ of dough on bottom and sides of pie plate evenly. Add filling. Dot with butter. Crumble the remaining ⅓ of dough on top. Bake at 400 degrees for 50 minutes.

Nectarine conserve

Almonds and raisins join nectarines in this flavorful conserve spiced with cinnamon.

4½ cups prepared fruit
½ cup slivered blanched almonds
½ cup seedless raisins
¼ teaspoon grated lemon rind
2 tablespoons lemon juice
½ teaspoon cinnamon
6 cups sugar
1 1¾-ounce package powdered pectin

Pit about 3½ pounds nectarines. Grind or finely chop. Measure 4½ cups into a 6- to 8-quart kettle. Add almonds, raisins, lemon rind and juice and cinnamon. Measure sugar and set aside. Mix pectin into fruit in kettle. Place over high heat and stir until mixture comes to a full boil. Add sugar and stir. Bring to full rolling boil and boil hard for 1 minute, stirring constantly. Remove from heat and skim off foam with metal spoon. Ladle quickly into hot jars, filling to within ¼ inch of top. Cover and process in boiling water bath for 5 minutes. Makes 8 or 9 ½-pint jars.

Nectarine care

To prevent discoloration after they are peeled or cut, dip nectarines in or sprinkle them with citrus juice (lemon, orange, lime or grapefruit). Serve at room temperature.

Peeling the thin, tender, smooth skin isn't necessary. But if you do want to remove the skin, submerge the fruit in boiling water for about 30 seconds; remove from water with a slotted spoon and plunge immediately into cold water. The skin will slip right off.

Olives

Botanical name: *Olea europaea*

There should have been an award for the first person who sampled the fruit of the olive tree and then had the persistence to develop a way to turn its bitterness into an edible delicacy.

The Bible mentions this ancient tree many times. In the Mediterranean area, for centuries a favorable spot for growing olives, the tree has long been valued for both the olives and the golden oil that can be extracted from them.

Olive trees were introduced to this part of the world by the Spanish and Italian missionaries.

Nowadays, their gray, driftwoodlike trunks and their slender silver-green leaves are a common sight in urban areas. They ornament the grounds of many public buildings, those on the University of Arizona campus being especially handsome.

Wherever they are grown, their charm is so great that no one gives them up, despite the problems their spring pollen brings to allergy sufferers.

Drought-resistant, olive trees nonetheless produce better when watered as other evergreen trees. They live to be hundreds of years old. Home growers can successfully debitter the green fruit for pickling, or can create the somewhat acrid but spicy Greek-style olive with ripe ones. Popular varieties are the Manzanilla, Mission and Sevillano.

Preparing olives for eating is not so much a difficult process as one that requires patience. Many a quart can be obtained from one female tree. (There needs to be a male tree nearby for fertilization.)

Ripe olives contain a good bit of oil — about 14 grams per 100 — but the proportion of unsaturated fats is higher than in any other fruit. Because they are preserved by salting and pickling, they are very high in sodium. The fruit can be pressed to make oil, but this is a difficult process for the home grower.

Brining olives

Jack Brown of Tucson prepares green or ripe-green olives using a non-fermenting procedure recommended by the University of Arizona Cooperative Extension Service some years back. The brined olives, sealed in jars, will keep for several months. They can be spiced or left plain.

Step 1. Pick, sort and wash olives.

Step 2. Pack olives lightly in large crocks or wide-mouth plastic or glass jars about two-thirds full. Cover with lye solution — 3 tablespoons of flake lye to 1 gallon of water. (A 13-ounce jar of lye costs about $1 at the grocery store.) Cover crock or jar with cloth or lid (not screwed on tightly), leaving in solution for three days. Stir from time to time with a large wooden spoon. Test for bitterness by slicing an olive to the pit, washing thoroughly and tasting a small piece of the fruit. If still bitter, give the olives two more days in the lye solution.

Step 3. When a test olive is free of bitterness, pour the lye water down the drain and wash the olives in clear water 2 or 3 times a day for a week.

Step 4. Now start the brining solution. First, use 4 ounces of salt to 1 gallon of water — pickling or ice-cream salt are cheaper than table salt, Brown said. Cover olives and leave in cool place for three days, stirring every day. Pour off brine.

Step 5. Add a second solution of 8 ounces of salt to 1 gallon of water, and leave for five or six days, stirring every day. Pour off brine.

Step 6. Transfer olives to quart jars and add a third solution of 14 ounces of salt to 1 gallon of water. Also add, if desired, five cloves of garlic, two teaspoons pickling spice, ½ teaspoon oregano or sprig of dill weed (or any combination desired.) Close lids tightly and store olives in cool, dark place.

Olives prepared this way will keep about six months if unopened, said Brown.

Using brined olives

Thelma Riggs of Green Valley, whose father used to process olives at their California home using a lye-and-brine method similar to the one given here, offers this reminder: Before being eaten, olives prepared this way should be taken from the brine and freshened in cold water for 24 hours. Keep the freshened olives in the refrigerator.

Olive oil

Making olive oil at home is a detailed process that is not very satisfactory, but it can be done. Details are given in bulletins from the University of California Cooperative Extension Service, Berkeley.

Greek olives

Mary Gekas *of Tucson uses this recipe when preparing spicy olives. Her method requires no lye.*

Olives should be picked when dark and ripe, but not overripe, as they bruise easily. Slash each olive deeply on each side with a knife and place olives in large non-metal containers. Pour a solution of salty water (4 tablespoons salt to 1 quart of water) over olives. Leave the olives uncovered and in a cool place.

Stir a little each day. Change the brine each week. Generally it requires 3 to 4 weeks to remove the bitterness of the olives. Sample after the third week, and continue the process if still bitter.

When they are no longer bitter, rinse olives well. Fill screw-top quart jars ⅔ full of olives. For each quart, prepare a mixture of 1 pint water, 1 pint vinegar, 1 tablespoon salt, ½ teaspoon oregano, 2 small lemon wedges and 1 clove garlic and pour over olives. Top with a ½-inch layer of oil and screw on lid tightly. Invert the jar for a week so the oil will seep through the olives. Store right-side up in cabinet. Olives will keep for several years.

Peaches

Botanical name: *Prunus persica*

Lovely to look at, fragrant and absolutely delicious to eat. Today's peach lovers would agree with that assessment.

The ancient Chinese also esteemed the peach and not only cultivated the fruit, but considered it a symbol of long life and immortality. When the peach moved westward along caravan routes to Asia Minor, it found a compatible home in the gardens of Persia, where it became known as the "Persian apple."

Eventually, travelers and traders carried it to Europe and on to the New World. Since the early 19th century, it has become one of our most important commercial fruit crops.

Though the peach grows well in cooler areas, it succeeds in the desert if the right varieties are chosen and proper care is given. Several desert varieties are being developed by the Israelis that show promise for low-chilling areas.

Meanwhile, some of the early ripening varieties popular for home orchards in Maricopa, Yuma and warmer areas of Pinal and Pima counties include Desert Gold, Blazing Gold and Junegold. In Yuma County, McRed bears good crops. Later in the season, the Tucson area is successful with Loring and Red Haven. Higher elevations find success with them as well, along with Delicious Peach, Babock and Kim Elberta.

Peaches are one of the most healthful of foods. Among their pluses are good quantities of vitamin A and potassium, along with other minerals. A medium fresh peach contains about 150 calories. Any family that dries, cans or preserves the fruit is assured eating delights all year.

Fried peaches

*Tucsonan **John Tanner** finds this peach dish simple to make and very good to eat.*

Wash and halve fresh peaches. Melt butter in frying pan and brown peaches on outside and then turn. Into each center, place 1 teaspoon brown sugar. Cover pan and simmer until peaches are tender, adding a little water if needed. (Canned peaches may be drained well and prepared the same way.) This dish is especially good at breakfast with bacon, sausage or ham.

Fresh peach dumplings

*One of the great dishes to be made with freshly picked peaches is Tucsonan **Hazel M. Battiste's** dumplings, but no peeping while they cook!*

3½ cups sliced peaches
2¼ cups water
¼ teaspoon nutmeg
1½ to 2 cups sugar
1 tablespoon lemon juice
1¼ cups biscuit mix
¼ teaspoon nutmeg
½ cup brown sugar
2 tablespoons oil
½ cup plus 1 tablespoon milk

Combine peaches, water, nutmeg, 1½ to 2 cups sugar and lemon juice and simmer in saucepan for 5 minutes. Mix well the biscuit mix, nutmeg and brown sugar. Add oil and milk and stir gently. Drop mixture by tablespoonfuls into hot peaches. Cook for 20 minutes, tightly covered. Serve warm with vanilla ice cream or topping of your choice. Makes 6 or 7 generous servings.

Peach kuchen

*A favorite for Sunday brunch, says Phoenician **Susanne S. Cheung**, is this kuchen.*

Batter
1½ cups sifted all-purpose flour
½ cup sugar
2 teaspoons baking powder
½ teaspoon salt
¼ teaspoon ground cardamom
2 eggs
2 tablespoons milk
1½ tablespoons grated lemon peel
¼ cup butter, melted

Peach filling
2½ pounds peaches
2 tablespoons lemon juice

Topping
¼ cup sugar
½ teaspoon ground cinnamon
1 egg yolk
3 tablespoons heavy cream
Sweetened whipped cream or soft vanilla ice cream

Cover peaches with boiling water in large bowl. Let stand for 1 minute to loosen skins. Drain and plunge in cold water for 30 seconds. Slice peaches and sprinkle with lemon juice. Set aside.

Sift flour with sugar, baking powder, salt and ground cardamom. In large mixing bowl, beat eggs with milk and lemon peel. Add flour mixture and melted butter; mix until smooth. Butter a 9-inch springform pan. Turn batter into pan. Spread evenly over bottom.

Combine sugar and cinnamon. Drain peaches. Arrange on batter in circular pattern. Sprinkle evenly with cinnamon-sugar mixture. Bake for 25 minutes. Beat egg yolk and heavy cream. Pour over peaches. Bake for 10 minutes longer. Cool for 10 minutes on wire rack. To serve, remove side of springform pan. Serve kuchen warm with sweetened whipped cream or soft vanilla ice cream.

Peach and nut cake

Summer's peaches go into this delightful upside-down cake, which can be made in a jiffy with a mix, if you prefer.

2 cups firmly packed brown sugar, divided
1 cup chopped pecans
2 cups peeled or unpeeled chopped peaches
¼ pound butter or margarine, softened
1 egg
1 cup milk
1 teaspoon vanilla
2 cups flour
1 teaspoon baking powder
½ teaspoon salt
½ teaspoon each cinnamon and nutmeg

Melt ¼ cup butter in 9- or 10-inch skillet. Toss together 1 cup brown sugar, pecans and peaches. Spoon over butter in pan. Cream together remaining sugar and butter. Add egg, milk and vanilla and stir until smooth. Sift together flour, baking powder, salt and spices. Add to egg mixture and stir. Spoon into prepared pan. Bake at 350 degrees for ½ hour, or until cake tester comes out clean. Allow to cool in pan 5 minutes, then gently turn out onto plate and cool. Serve warm or cold.

Note: If using cake mix, prepare ¼ cup butter, 1 cup brown sugar, 1 cup pecans and 2 cups peaches, as described above, and top with cake batter from a mix.

Fresh peach cake

Bea Pease *of Fort Huachuca adapted this recipe from a plum cake recipe in an effort to duplicate the fresh peach cake that was popular in Baltimore when she lived there years ago.*

6 large peaches, peeled
1 package dry yeast
1 cup lukewarm milk
3½ cups flour
1 teaspoon salt
1 cup sugar
¼ cup butter
1 slightly beaten egg
2 tablespoons butter
½ teaspoon cinnamon

Wash and peel peaches, remove pits. Sprinkle yeast into ¼ cup of milk. Let stand until dissolved. Mix flour, salt and ⅓ cup sugar in large bowl. Stir dissolved yeast, ¼ cup butter, egg and rest of milk into dry ingredients. Knead to make a soft dough. Let rise 30 to 40 minutes. Punch down and roll out on floured surface into a 9-by-13-inch oblong. Fit dough into greased 9-by-13-inch pan. Flute edges to make a slight rim. Let rise for 20 minutes more. Put peaches, cut side up, on top of dough. Sprinkle with rest of sugar mixed with the cinnamon. Dot with 2 tablespoons butter. Bake at 350 degrees for 35 to 40 minutes. Serve warm or cold, with or without French vanilla ice cream.

Peach ice cream

Fresh peaches and sweetened condensed milk are combined for a delicious ice cream in this recipe from **Dianna Dameron** *of Thatcher.*

2 3-ounce packages lemon-flavored gelatin
1 15-ounce can sweetened condensed milk
1 cup sugar
1½ cups chopped fresh peaches
Milk as needed

Mix the gelatin, sweetened condensed milk and sugar in an ice cream freezer-can. Then mix in the peaches and add enough extra milk to bring mixture to within 2 inches of the top. Freeze according to freezer manufacturer's directions. Makes 2 quarts.

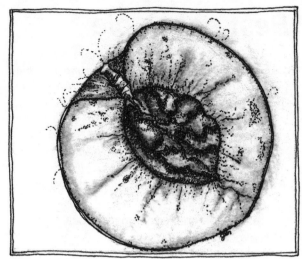

Peach

Fresh peach ice cream

When fresh peaches are in, it's time to make ice cream.

1 cup heavy cream
2 cups milk
4 egg yolks
1 cup sugar
  Salt to taste
1 teaspoon pure vanilla extract
1 cup ripe peaches, peeled,
  pitted and crushed

In a saucepan with a heavy bottom combine the cream, milk, egg yolks, half a cup of the sugar, salt and vanilla extract. Cook over low heat or in a double boiler, stirring constantly with a wooden spoon all around the bottom to make sure the custard does not stick. Continue cooking and stirring until the custard is as thick as heavy cream (180 degrees). Remove the custard from the heat immediately, stirring constantly for a minute or so. Let cool.

To the peaches add the remaining half a cup of sugar. Stir to dissolve. Add this to the custard. Pour the mixture into the container of an electric or hand-cranked ice cream freezer and freeze. Makes 8 to 10 servings.

Peach sherbet

Peach sherbet is just the thing for a lift when fresh peaches are in.

½ cup sugar
⅓ cup light corn syrup
⅓ cup lemon juice
¾ to 1 pound fresh peaches
2 cups milk

In medium bowl stir together the sugar, corn syrup and lemon juice. Peel, pit and slice enough of the peaches to make 2 cups; purée in electric blender (1½ cups are required). Add the purée and milk to the sugar mixture. Stir until sugar dissolves. Pack into 1 or 2 appropriate containers and freeze. Sherbet becomes icy-hard and needs to stand at room temperature for 10 to 15 minutes to soften (or soften in microwave oven). Makes 1¼ quarts.

Peachy pudding

This up-to-date version of the old Grape Nuts dessert from **Lisa Goode** *of Nogales substitutes wheat germ for the cereal.*

1 egg
2 tablespoons sugar
1 teaspoon flour
  Pinch of salt
1 cup drained canned peaches
¼ cup wheat germ
¼ teaspoon vanilla
1 teaspoon sugar

Beat egg slightly; beat in 2 tablespoons sugar. Blend in flour and salt. Fold in peaches, wheat germ and vanilla. Spoon into greased 1¾-cup baking dish (or bake in 2 greased 5-ounce custard cups); sprinkle with remaining sugar; and cover with aluminum foil. Bake for 15 minutes at 350 degrees. Uncover and bake for additional 10 minutes, or until lightly browned. Serve warm with ice cream, if desired. Serves 2.

Peach Melba split

Try the old favorite, peach Melba, in this updated version using bananas and raspberry sherbet.

2 tablespoons sugar
2 cups sliced peaches
4 bananas
1 pint raspberry sherbet

In medium bowl sprinkle sugar over peaches. Let stand 15 minutes. Peel bananas, cut in half lengthwise, and place 2 halves on each dessert plate. Scoop raspberry sherbet over bananas. Top with peaches. Serves 4.

Peach nectar

Fruit nectar is a delicious way to use an abundant fruit crop.

8 pounds ripe peaches
4 teaspoons each lemon and salt
Water
½ cup lemon juice
1 cup sugar

Dip peaches in boiling water to loosen skins. Dip in cold water and remove skins, halve and pit. Drop pieces into a mixture of the 4 teaspoons lemon juice, 4 teaspoons salt and 2 quarts water. Drain and cut into smaller pieces and put in electric blender or food processor in amounts that each will hold. Purée.

Pour purée into a large kettle and add the ½ cup lemon juice, the sugar and 10 cups water. Heat and stir occasionally until mixture reaches 165 degrees on a candy thermometer. Remove from heat, sample and add additional sugar, if desired. Skim off foam and ladle nectar into clean jars or decanters. Seal and process in boiling-water bath for 20 minutes. Makes about 5 quarts or 6 26-ounce canning decanter jars.

Note: Nectar may be diluted at serving time, if desired.

Best peach pie

When the peaches are ripe, that's when **Mamie Grizzle** *of Elfrida makes her memorable pies. A little vanilla in the filling is a flavorful touch.*

Pastry for 9-inch, 2-crust pie
About 5 cups peaches, peeled and sliced
2 tablespoons butter
¾ to 1 cup sugar, depending on sweetness of fruit
1 tablespoon quick-cooking tapioca
1 teaspoon vanilla

Line pie pan with pastry and dot with butter. Layer with peaches. Combine tapioca with sugar and sprinkle over peaches. Last, sprinkle filling with vanilla. Add top crust, slash and bake at 400 degrees for about 50 minutes. Makes 1 pie. Recipe may be doubled or tripled as desired. Pies may be baked and frozen for use later. Defrost and reheat.

Super peach pie

This fresh peach pie has won first-prize ribbons for **Jolene Randall** *of Tucson.*

5 large peaches
1 cup sugar
3 tablespoons cornstarch
½ cup water
½ teaspoon almond flavoring
2 tablespoons butter
1 baked pie shell
Whipped cream

Slice 3 of the peaches into the pastry shell. Mash remaining peaches and cook in saucepan with sugar, cornstarch and water, stirring, until it boils (about 5 minutes). Add almond flavoring and butter. Pour over sliced peaches. Chill. Serve with whipped cream.

All-season peach pie

Hester Irwin of Nogales calls her pie "all-season" because when there are no fresh peaches, canned ones will do.

1 unbaked 9-inch pie shell
   Peach halves to cover bottom of pie shell
½ cup sugar
3 tablespoons cornstarch
¾ teaspoon nutmeg
¼ teaspoon salt
¾ cup heavy cream
¾ teaspoon vanilla
   Whipped cream

Arrange peach halves, cut side up, in pie shell. Combine sugar, cornstarch, ½ teaspoon nutmeg, salt, cream and vanilla. Pour over peaches. Sprinkle with remaining nutmeg. Bake at 400 degrees for 40 minutes or until peaches are tender. Serve with whipped cream. Makes 1 pie.

Soda cracker pie

Tucsonan **Agnes Franklin** got this recipe from a college friend and finds it simple to make and delicious.

3 egg whites
1 cup sugar
14 soda crackers
¼ teaspoon baking soda
⅓ cup chopped pecans
1 teaspoon vanilla
   Peaches, peeled and sliced, to layer in pie plate
1 cup whipping cream

Beat egg whites until stiff and gradually beat in sugar. Roll crackers into very fine crumbs and add baking soda. Fold into egg whites. Add pecans and vanilla. Bake in buttered pie plate at 325 degrees for 30 minutes. Cool. Top with layers of sliced peaches. Store, covered in refrigerator. Just before serving, whip cream and spread on top.

Peach glacé pie

This pie originally was made with strawberries, but **Helen Dotson** of Tucson changed it to peaches and finds it especially nice to make with her food processor.

1 9-inch baked pastry shell
4 cups sliced, peeled peaches
3 to 4 ounces cream cheese, softened
⅔ to ¾ cup sugar
3½ tablespoons cornstarch
   Whipped cream (optional)

Mash, blenderize or process in food processor 2 cups sliced peaches. If necessary, add water to make 1½ cups. Gradually stir in mixture of sugar and cornstarch. Cook over low heat, stirring constantly until boiling. Boil 1 minute. Cool. Add remaining 2 cups sliced peaches and stir gently to mix. Spread softened cream cheese over bottom of cooled baked pastry shell. Pour peach mixture into cream cheese-lined pie shell and chill about 2 hours. Before serving, garnish with whipped cream, if desired.

Peach cobbler

Mary Shields of Marana makes many a peach pie from her husband's orchard. Make more than one and freeze before baking.

Fill a rectangular, shallow casserole (9-by-13 inches) with peeled sliced peaches, add ¾ cup sugar, 2 rounded tablespoons cornstarch and a good dash of nutmeg and cinnamon. Stir. Dot with 1½ tablespoons butter. Cover with strips of pastry, using ½ a standard pastry recipe. Bake 10 to 15 minutes at 425 degrees, lower temperature to 350 degrees and bake about 20 minutes more, until crust is browned and peaches are tender.

Spiced peaches with almonds

Medium-size peaches, along with slivered almonds and raisins, go into these well-seasoned jars, along with a touch of rum.

2 cups sugar
1 cup raisins
5 to 6 cinnamon sticks
3 cups water
¼ cup light rum
½ teaspoon whole cloves
3 whole allspice berries
3 quarts peeled and pitted fresh peach
    halves (10 to 12 medium-size peaches)
½ cup slivered blanched almonds

In a large saucepot, combine sugar, raisins, cinnamon sticks, water and rum. Tie cloves and allspice in cheesecloth; place in saucepot. Stir mixture until sugar is dissolved. Add peaches and almonds. Bring to a boil. Boil for 5 minutes, stirring frequently. Pack peaches and syrup into canning jars, leaving ¼-inch head space.

Seal and process in boiling water bath for 15 minutes. Cool jars and check seals. Makes 5 to 6 pints.

Peach and marshmallow cake

This recipe was her mother's and using it has given much eating enjoyment to Tucsonan **Florence E. Horner.**

1 tablespoon gelatin
¼ cup cold water
⅓ cup butter
1 cup powdered sugar
32 marshmallows
2 eggs, separated
4 cups sliced peaches
2 cups crushed vanilla wafers

Soften gelatin in cold water for 5 minutes. Cream butter, add powdered sugar and blend in egg yolks. Cook over low heat, stirring constantly until thickened. Remove from heat and add softened gelatin. When slightly cool, add marshmallows cut in small pieces. Blend lightly and cool. Fold in peaches and stiffly beaten egg whites.

Beginning and ending with crumbs, put alternate layers of crumbs and peach filling in a mold. Chill until firm. Unmold and serve with or without whipped cream. Serves 8.

Fresh peach cobbler

Connie Young *of Winkelman shares this cobbler recipe, delicious also with apples or cherries.*

Sliced fresh peaches (about 3 cups)
Sugar for peaches
1 cup flour
1 cup sugar for pastry
1 teaspoon baking powder
¼ teaspoon salt
1 beaten egg
½ cup evaporated milk
⅓ cup cooking oil
3 tablespoons chopped pecans
Cinnamon and sugar

Arrange peaches in greased 7-by-9-inch pan and fill with 2 layers of fruit. Add sugar to taste. Mix flour, sugar, baking powder and salt together. Combine egg, milk and oil and then combine with first mixture. Pour flour mixture over the fruit. Sprinkle top with pecans, cinnamon and sugar. Bake at 350 degrees for 50 minutes. Serves 8.

Pears

Botanical name: *Pyrus communis*

An ancient fruit from Asia, the pear was greatly admired by the Greeks. In Rome's heyday, there were 41 species of the fruit growing throughout the empire, and by the time the colonists and Spanish mission fathers brought them to the New World, there were thousands of varieties cultivated in Europe.

In Colonial days, pears were valued almost on a par with apples. The settlers ate the pears, made furniture from the wood and yellow dye from the leaves.

Nowadays, Western commercial growers limit their production mainly to Bartletts (the best known), Comice, Bosc and d'Anjou, although many others are regionally popular.

Pears make attractive, stately trees for the yard. Dwarf varieties can be espaliered, and both full-size and dwarfs provide delicate white flowers that are among spring's glories.

In Arizona, pear trees generally will not bear fruit below 1,200 feet. Best choices are Bartlett and Keifer. From 3,500 feet to 6,000 feet, Bartlett, d'Anjou, Seckel and Comice are recommended.

Though it is lower in vitamin A than the apple, the pear otherwise resembles the apple in nutrients. A medium-size pear contains about 100 calories, is high in fiber and low in sodium, and offers both B vitamins and trace minerals.

Pears are among the most delectable of fruits when dried, but also make excellent additions to the larder when canned or preserved.

Chicken with pears and lemon

Fresh pears and grated lemon peel make this skillet chicken a memorable main dish.

2½ to 3-pound broiler-fryer, cut up
½ lemon
 Salt
¼ cup butter or margarine
½ teaspoon grated lemon peel
¼ cup sherry or chicken stock
 2 fresh pears, cored and sliced
 1 or 2 teaspoons flour
¼ to ½ cup water
 Paprika

Rub chicken with lemon, squeezing juice onto chicken; sprinkle lightly with salt. Let stand 15 minutes.

In skillet, brown chicken well in butter or margarine. Drain excess fat; pour sherry over chicken. Sprinkle with lemon peel. Simmer, covered, 30 to 40 minutes or until chicken is tender, turning once. Add pears, and simmer, covered, for an additional 5 to 10 minutes or until pears are cooked, but still firm. Remove chicken and pears to warm serving platter. Blend flour into drippings; cook until thickened. Add water to desired consistency. Spoon sauce over chicken and fruit; sprinkle with paprika. Serves 3 or 4.

Fresh pear cake

Laurel Collier Decker of Tucson shares this interesting-sounding spiced pear cake recipe, which comes from her mother in Oregon.

3 cups flour
2 teaspoons baking powder
2 teaspoons baking soda
1½ teaspoons cinnamon
2 cups sugar
4 eggs
1½ cups oil
1¼ cups peeled and mashed fresh pears
1½ cups chopped almonds

Combine dry ingredients. Add oil and eggs; fold in pears and almonds. Pour into well-greased tube pan and bake at 350 degrees for 1 hour or more. Cool before turning out.

Poached pears

The mellow flavor of tree-ripened pears makes this dessert a rare treat.

½ cup sugar
¾ cup white wine
¼ cup sherry
½ cinnamon stick
2 to 3 whole cloves
⅛ teaspoon powdered ginger
2 ripe pears, peeled, cored and halved

Combine sugar, wine, sherry, cinnamon, cloves and ginger in medium-size saucepan. Bring to boil, reduce heat and simmer 5 minutes. Meanwhile, prepare pears. Add pears to saucepan and simmer until pears are fork tender. Remove pears from syrup. Bring syrup to boil again and boil about 5 minutes, or until syrup is reduced and slightly thickened. Pour syrup over pears and chill for 1 hour or more. Makes 2 servings.

Glazed pear pizza

This recipe makes a hit wherever Elaine Drorbaugh of Payson takes it. She obtained it from an Oregon relative during a family camp out.

1 recipe sugar cookies or 1 tube
   sugar-cookie dough
1 8-ounce package cream cheese
1 cup powdered sugar
3 to 4 cups canned pears
   Juice from canned pears plus water
   to make 1 cup
2 tablespoons cornstarch

Press dough on pizza pan or cookie sheet. Make a rim around the edge. Bake until brown and cool. Combine cream cheese and powdered sugar. Spread on baked cookie dough. Then cover cheese with well-drained fruit. Add drained liquid plus water to make 1 cup. Stir in cornstarch. Cook and stir over medium heat until thickened. Pour over fruit.

Note: Other canned fruit or fresh fruit may be used, if desired.

Pear syrup

This very old recipe from Tucsonan Connie Scharpnick's grandmother is great on toast or over pancakes.

4 pounds cut and peeled pears
   (about 7)
4 cups sugar
1 cup water

Cut pears in thin slices. Bring sugar and water to a boil in a large pot. Stir and add pears and cook until pears are nearly clear. The liquid will be thin. Pour into clean jars and seal.

Pirtle's pear pie

Klaire Pirtle *of Tucson received this recipe years ago through a recipe chain letter, and found it a yummy dessert.*

Pastry for 9-inch single-crust pie
1 tablespoon fine, dry, white bread crumbs
¼ cup sugar
⅓ teaspoon ginger
4 teaspoons flour
4 teaspoons lemon juice
5 cups cored and thinly sliced ripe
    Bartlett pears (about 1⅓ pounds)
¼ cup white corn syrup
    Streusel topping

Line pie pan with pastry and flute edges or mark with a fork. Sprinkle crumbs over pastry. Blend sugar, ginger and flour and sprinkle ⅓ over the crumbs. Spread pear slices evenly in pan, drizzle with lemon juice and syrup. Sprinkle with remaining sugar mixture and top with streusel. Bake at 450 degrees for 15 minutes, reduce heat to 350 and bake for 30 to 35 minutes longer. Cool to lukewarm before serving.

Streusel: Blend ⅔ cup flour and ⅓ cup sugar in 2-quart bowl. Add ⅓ cup butter and cut to pea-size particles with pastry blender or 2 knives.

Pear conserve

Pears combine with apples and orange for a distinctive and delicious conserve in this recipe from **Janet Black** *of Douglas.*

6 large pears
2 green apples
1 orange
10 maraschino cherries
3 cups sugar

Wash fruit. Remove seeds and core but do not remove skins. Put through medium grinder of food chopper, or food processor. Add sugar. Bring to a boil, stirring constantly. Boil for 15 minutes. Pour into ½-pint sterilized jars and seal. Makes about 4 ½-pints.

Pear butter

Dorothea Jackson *of Tucson sends this recipe for pear butter. The same recipe may be used for apples, peaches, apricots, etc.*

Wash fruit thoroughly and cut in pieces, but do not remove skins or seeds. Cover with cold water. Bring to boil, then simmer until soft. Rub through coarse sieve. Measure pulp. To each quart allow 2 cups sugar, 1 teaspoon ground cinnamon and ½ teaspoon ground cloves. Return to heat and simmer until thick. Turn into hot sterilized jars and seal immediately.

Persimmons

Botanical names: *Diospyros kaki (Japanese); D. virginiana (native)*

Sour as can be, unless fully ripe, persimmons are a round, bright orange-colored fruit about 2 inches in diameter. They come in two species: American and Japanese.

Persimmon trees are slender and can grow to 40 feet in height. Though especially popular in the South, they can be grown in the southern half of Arizona. At altitudes between 1,200 to 3,500 feet, the recommended varieties are Hachiya and Fuyu, both Japanese types. The fruit ripens in the fall.

Persimmons are especially tasty in baked goods. The seeds (there are four to eight of them) are sometimes roasted and used as a coffee substitute. The fruit is high in pectin and is considered an outstanding source of vitamin A. It also is high in potassium.

The fruit was a special delight of Capt. John Smith of Jamestown, who once wrote that a ripe persimmon "is as delicious as an apricock."

Persimmons are not important commercially, though small numbers of them are grown in California.

Persimmon salad

Try this salad with a lime and honey dressing for an unusual taste sensation.

Mixed salad greens for 6
4 ripe but firm persimmons
2 tablespoons lime juice
1 tablespoon honey
⅓ cup salad oil
Dash salt and cinnamon

Peel and slice persimmons and remove seeds. Mix with salad greens. Beat lime juice, honey, salad oil, salt and cinnamon until well blended. Pour over salad and mix gently. Serves 6.

To prepare

Persimmons may be eaten out of hand when ripe. Simply wash them, cut off the flower end and cut in half. Eat with a spoon or fork as a dessert course.

Or, place flower-end-down on a serving plate and serve with a pointed knife and a pointed spoon. To eat, slit the skin in sections, pull the sections back and spoon up the pulp.

The pulp may be puréed and frozen for use when the fresh fruit is unavailable.

Persimmon spice cake

Use puréed persimmon, fresh or frozen, in this recipe.

1¼ cups persimmon purée
1 teaspoon soda
½ cup butter or margarine
1 cup sugar
2 eggs
2 cups flour
2 teaspoons baking powder
¼ teaspoon salt
1 teaspoon cinnamon
½ teaspoon ground cloves
½ teaspoon nutmeg
½ cup chopped almonds
1 teaspoon grated lemon peel
1 teaspoon grated orange peel
Powdered sugar

Stir purée and soda until well-blended; let stand while creaming butter and sugar. Add eggs. Add persimmon mixture to creamed mixture and blend well. Sift flour with baking powder, salt and spices. Gradually blend into batter. Add almonds and fruit peels. Beat until just well-blended. Spoon batter into greased and floured 9-inch square baking pan or a 5-cup mold. Bake at 350 degrees for 40 to 50 minutes. Let cake cool in pan for 10 minutes, then invert onto cooling rack for 15 minutes. Serve warm, dusted with powdered sugar.

Persimmon nut cake

Persimmons give this cake a rich golden color.

½ cup butter or shortening
1 cup sugar
1 egg
1¾ cups sifted flour
1 teaspoon soda
¼ teaspoon salt
1 teaspoon cinnamon
½ teaspoon mace
1 cup persimmon pulp
1¼ cups raisins or dried currants
1 cup chopped nuts

Cream butter and sugar until fluffy. Add egg and beat well. Sift flour again with soda, salt and spices and add to creamed mixture alternately with persimmon. Stir in lightly floured raisins and nuts. Line bottom of a 9-by-5-inch loaf pan with waxed paper and pour in batter. Bake at 350 degrees for about 1 hour. Served plain or topped with butter frosting. Makes 1 loaf.

Persimmon cookies

Spices, nuts and raisins make these moist persimmon cookies from **Betty Payne** *of Mesa extra special.*

1 teaspoon soda
1 cup persimmon pulp
½ cup butter or margarine
1 cup sugar
1 egg
2 cups flour
½ teaspoon salt
½ teaspoon each cinnamon, cloves, nutmeg
1 teaspoon baking powder
1 cup raisins
1 cup nuts (optional)

Peel persimmons and press through a colander. Mix soda with persimmon pulp and set aside. Cream butter or margarine and sugar. Beat in egg and add persimmon-soda mixture. Sift dry ingredients together and add to first mixture. Mix in raisins and nuts. Drop by spoonfuls on a greased cookie sheet and bake at 350 degrees for 8 to 10 minutes or until done. Makes about 4 dozen cookies.

Persimmon pie

Sandy Migliacci *of Tucson shares this recipe that was given to her years ago by her mother.*

2 cups persimmon pulp
½ cup sugar
½ teaspoon mace
1 teaspoon grated lemon rind
⅛ teaspoon salt
2 teaspoons butter
2 eggs, separated
1 baked pastry shell
4 tablespoons sugar
½ teaspoon vanilla

Persimmons should be sweet and very ripe. Peel and press enough of the fruit through a colander to make 2 cups of pulp. Add sugar, mace, lemon rind and salt, and cook slowly for 5 minutes. Beat egg yolks and add butter. Stir small amount of hot persimmon mixture into egg yolks and butter. Stir yolk mixture into hot persimmon mixture and cook and stir until mixture is slightly thickened. Pour into pastry shell; cool. Beat egg whites until peaks form, gradually adding sugar while beating. Continue beating until stiff. Add vanilla and pile on pie. Bake at 325 degrees until lightly browned.

Pineapple guavas

Botanical name: *Feijoa sellowiana*

This fruit grows on an evergreen shrub or small tree that requires little water, although it can take lawn water. Natives of South America, pineapple guavas may be trained to decorate a wall or they can be clipped into a hedge. However, if this is done, fruit production is lowered. It can be grown in areas suitable to citrus, but is more likely to bear fruit in the midaltitude desert than the low desert. The plant is not the same family as the strawberry guava or lemon guava *(Psidium species)*.

The glossy leaves of the pineapple guava are 2 to 3 inches long. They are a distinctive gray-green on top and silvery-white underneath. The inch-wide spring blossoms have four fleshy white petals and deep rose-colored stamens. The petals, minus the centers, are edible and attractive in fruit salads.

The fall-ripening fruit, an inch or so in length, is a light yellowish green, with yellowish pulp inside. When ripe, it smells very much like a pineapple. It is good eaten fresh or made into jelly. Popular varieties are Coolidge and Pineapple Gem.

Plums

Botanical name: *Prunus domestica and P. salicina*

Delicious as they are, plums don't seem to have inspired romantic stories as some fruits have done. The best known story about plums in English-speaking countries is likely Little Jack Horner's experience. The nursery rhyme refers to a scandal in the time of Henry VIII, when he was taking over properties belonging to the Catholic Church. Horner was a steward assigned by one of the abbots to carry a Christmas gift to the king: a pie in which deeds to several estates were hidden. The sticky-fingered Horner opened the pie and pulled out his plum — a deed to one of the manors. All of which explains why the term "plum" refers to something choice.

As for the fruit, plums have been cultivated for more than 2,000 years in the Orient. They were brought to this country from Europe in the 17th century, and from Japan about 100 years ago. Two types of plums are grown here: European (blue to purple in color) and Japanese (red or yellow), and both thrive in the Sonoran Desert and adjoining areas.

Highly recommended for growing at elevations up to 3,500 feet are the Santa Rosa, a red plum of high quality; Beauty, an early ripening, amber-fleshed plum, and Laroda, a yellow-fleshed plum ideal for drying.

Above that level up to 7,000 feet, the preferred fruits are Santa Rosa and Duarte, both red-skinned; and Satsuma, purple. Also suitable for this elevation is

Stanley Prune, a large, dark blue fruit. All require cross-pollination.

Arizona produces some plums commercially, but California produces a whopping 90 percent of the country's commercial supply.

Plums are low in calories, having only 30 calories in an average 3-ounce plum. They also are low in sodium, high in potassium and contain fair amounts of vitamin A. All plums dry beautifully. The purple ones, when dried, are called "prunes."

Barbecued chicken with plum sauce

An interesting change for the patio grill is this barbecued-chicken sauce made with fresh red plums.

  3 red plums
¼ cup brown sugar, packed
¼ cup tomato sauce
¼ cup sliced green onion
  2 tablespoons vinegar
  1 teaspoon Worcestershire sauce
¼ teaspoon salt
½ cup white wine
¼ cup finely chopped celery
¼ cup finely chopped green pepper
  Chicken pieces to serve 4

Pit and chop plums and purée in blender. Add sugar, tomato sauce, green onion, vinegar, Worcestershire sauce and salt. Blend smooth. Divide sauce in half and add wine to one portion for basting sauce.

Stir celery and green pepper into second portion to serve as relish. Place chicken pieces, bone side up, on grill (or in broiling pan of oven). Grill about 5 inches from heat for 15 minutes. Turn and brush with basting sauce. Continue to broil for 15 minutes, brushing often with basting sauce until chicken is done and nicely glazed. Serve with the relish. Serves 4.

Plum noodle kugel

If you've never tasted a fruit filled noodle kugel, it's high time. This version is quite good.

  6 ounces noodles, uncooked
  3 eggs
  1 cup sugar
  1 teaspoon cinnamon
3¾ cups fresh plums, quartered and pitted
  1 cup applesauce
  1 cup soft bread crumbs
½ cup chopped nuts
  2 tablespoons melted butter

Cook noodles in boiling salted water until barely tender. Drain and rinse in cold water. Beat eggs. Add sugar and cinnamon; mix well. Toss noodles in egg mixture; stir in plums and applesauce. Pour into buttered 2-quart casserole. Combine bread crumbs, nuts and butter; sprinkle over top.

Bake at 350 degrees for 50 minutes or until golden. Serve warm with heavy or whipped cream. Serves 6 to 8.

Plums

Fresh plum kuchen

Make this simple fruit kuchen with plums or other fresh fruits, suggests Tucsonan **Margarette Hughes.**

1 cup flour
1 tablespoon sugar
½ cup margarine, slightly softened
1 egg, separated
1 tablespoon milk
2 pounds fresh plums
1 whole egg
¾ cup sugar

Combine flour and 1 tablespoon sugar and work in margarine with a pastry blender or 2 forks. Beat egg yolk with milk and work into flour mixture. Place in 9-by-9-inch square pan, patting over bottom and up sides for 1 inch. Cut plums in half and remove seeds. Place in rows on pastry. Beat egg white and whole egg, adding ¾ cup sugar slowly. Pour over fruit. Bake at 350 degrees for about 50 minutes. Serve warm or cold. Serves 6 to 9.

Note: If desired, make dish with 2 pounds of other fruit, such as thick slices of apples, quarters of peaches or pears (all peeled) or halves of apricots (unpeeled).

Plum butter

A dash of several spices gives the right touch when making plum butter.

1½ pounds fresh plums, sliced and pitted
2 cups sugar
½ teaspoon cinnamon
Dash each: nutmeg, cloves and allspice

In heavy kettle, stir plums and ½ cup of the sugar over lowest heat. When juices flow, bring to boil; cook for 5 minutes or until fruit is tender. Pour into blender. Chop, on low speed, 2 seconds. Return mixture to kettle; add remaining sugar and spices. Bring to boil. Reduce heat, yet maintain active bubbling; cook for 5 minutes, stirring frequently. Butter is done when it sheets from a metal spoon.

Or, pour a spoonful on a chilled saucer. If no rim of liquid forms around edge of butter, it is done. Ladle at once into hot, sterilized screw-top jars, leaving ½-inch headspace. Seal. Makes 3 half-pints.

Note: For a thicker butter, cook 5 minutes more.

Plum coffeecake

Sliced fresh plums baked on top of coffeecake create a great Sunday breakfast dish.

Coffeecake:
2 cups buttermilk biscuit mix
½ cup sugar
½ cup water
2 cups sliced fresh plums

Orange sauce:
½ cup sugar
1 tablespoon cornstarch
¼ teaspoon salt
1 cup orange juice
1 tablespoon butter or margarine

In mixing bowl, combine biscuit mix, ¼ cup sugar and water; mix well. Pat dough into greased 8-inch square baking pan. Cover top of dough with sliced plums; sprinkle remaining ¼ cup sugar over top.

In small saucepan, blend together ½ cup sugar, cornstarch and salt. Add orange juice; cook over low heat, stirring constantly, until mixture is clear, about 5 minutes. Remove from heat; stir in butter. Pour ½ of sauce over top of plums and dough. Bake at 425 degrees for 30 minutes. Serve warm with remaining orange sauce. Serves 6.

Dutch plum cake

*Tucsonan **Kay Trondsen's** family is always delighted when she makes this plum dessert, whether made with Santa Rosa or purple prune-plums.*

1 cup sifted flour
1½ teaspoons baking powder
½ teaspoon salt
6 tablespoons sugar
¼ cup shortening
1 egg
¼ cup milk
5 or more plums, cut in eighths
1 teaspoon cinnamon
¼ teaspoon nutmeg
3 tablespoons melted butter or margarine
⅓ cup currant jelly
1 tablespoon hot water

Sift flour, baking powder, salt and 3 tablespoons sugar. With 2 knives or pastry blender, cut in shortening until mixture is like coarse cornmeal. With fork, stir in combined egg and milk. Spread dough in greased 12-by-8-by-2-inch baking dish. On top arrange plums, slightly overlapping in parallel rows. Sprinkle with combined cinnamon, nutmeg and the remaining 3 tablespoons sugar and butter. Bake for 35 minutes at 400 degrees or until plums are tender. Beat jelly with hot water and brush over fruit. Serve warm, cut in squares, with or without a vanilla sauce or ice cream.

Mexican plum pies

Though this isn't a traditional Mexican dish, using flour tortillas as the base for a fruit dessert is an appealing idea.

1 to 1½ cups sugar
2 pounds plums, halved and pitted
Butter or margarine, softened
8 flour tortillas, 7 to 8 inches in diameter
4 teaspoons sugar
1 teaspoon cinnamon
Grated semisweet chocoloate
Sour cream

Sprinkle sugar over plums in saucepan

and let stand until sugar dissolves and juices form. Bring to boil and simmer until tender, about 10 minutes. Cool. Butter both sides of tortillas lightly. Arrange on cookie sheet. Combine 4 teaspoons sugar and cinnamon and sprinkle over tortillas. Bake at 350 degrees for 15 minutes. Cool on rack. To serve, sprinkle some chocolate on each tortilla. Spoon stewed plums over it and top with sour cream. Serves 8.

Pomegranates

Botanical name: *Punica granatum*

"Eat the pomegranate, for it purges the system of envy and hatred," Mohammed advised his followers.

Whether pomegranate fruit works in this wondrous fashion may be open to question, but no one with pomegranate trees in the yard can fail to appreciate the decorative effect of the scarlet flowers of this ancient fruit tree in the spring, and the reddish gold fruit from late summer to fall.

The minuscule amounts of bright red pulp surrounding each of the pomegranate's numerous seeds are rich in vitamins B and C, and contain some vitamin A. In a medium-size pomegranate, the pulp measures about 90 calories. Advocates of herbal remedies maintain that pomegrantes have a cleansing and cooling effect on the system.

Pomegranate juice is the stuff of which the lovely sweetening syrup called grenadine is made, an indispensible ingredient for many drinks. The juice also makes the prettiest of jellies. The seeds and their bright envelopes become the edible garnish for salads and fruit cocktails.

When picking pomegranates, look for those with skin or rind that appears to be thin and tough. Cut one open, and if there is an abundance of bright red flesh, they are ready for juicing. Sometimes, when cutting into the fruit, a dark area is discovered. This is caused by bacteria that develop after a hole is made in the rind by fruit-loving insects. If not too widespread, the dark area can be cut away, and the remainder of the pulp can be used.

In some varieties, the fruit bursts wide open on approaching maturity, while in others it cracks. When fully ripe, pomegranates keep well in the refrigerator for several weeks.

There are dwarf forms of the pomegranate that do not produce fruit and are grown just for ornamental purposes.

For the home orchard in elevations between 1,200 and 3,500 feet, Yellow Papershell and Wonderful are the recommended varieties. The fruit also can be grown in warm microclimates of somewhat higher areas.

Waldorf salad with pomegranate seeds

Tina Poindexter of Safford shares this recipe from her mother, **Minnie Taylor** of Pima. The pomegranate seeds on top give it a special touch.

2 cups diced apples
1 cup diced celery
1 cup chopped pecans
1 cup raisins (optional)
   Dash salt
   Mayonnaise
   Pomegranate seeds

Combine apples, celery, pecans, raisins and salt and moisten with mayonnaise. Garnish with ripe red pomegranate seeds. Serves 6.

Pomegranate tossed salad

With about a dozen pomegranate trees in her yard, **Eileen Jones** *of Thatcher devises various ways to use the fruit.*

1 medium head lettuce,
   torn or chopped
1 medium cucumber, diced
1 large tomato, diced
1 large avocado, diced
2 tablespoons chopped onion
1 cup pomegranate seeds

Toss ingredients and top with favorite dressing. Serves 6.

Pomegranate-sauced chicken

Tucsonan **Carolyn Niethammer** *passes along this attractive-sounding main dish using pomegranate juice in the sauce.*

3 pounds favorite chicken parts
Oil
½ teaspoon poultry seasoning
1 large onion, chopped
1 cup coarsely chopped pecans
2 tablespoons tomato sauce
1½ cups water
1 teaspoon salt
½ teaspoon ground cinnamon
1 cup orange juice
1 cup pomegranate juice
1 tablespoon cornstarch
Pomegranate seeds

Brown chicken pieces in oil, remove from pan and keep warm. Drain all but 2 tablespoons drippings from pan and stir in poultry seasoning and onion. Cook until transparent. Stir in nuts and cook until lightly brown. Combine tomato sauce with 1¼ cups water, the salt, cinnamon, orange juice and pomegranate juice; bring to boil and simmer for 20 minutes. Return chicken pieces to pan, cover and simmer until tender. Remove chicken. Combine cornstarch with remaining ¼ cup water and stir into sauce. Cook and stir until thickened. Serve chicken topped with pomegranate sauce and garnished with pomegranate seeds. Serves 6.

Pomegranate dessert sauce

For serving over ice cream or other desserts, **Doris Schultz** *of Green Valley finds this sauce appealing.*

Combine ½ cup sugar and 1 tablespoon cornstarch; stir in 1 cup pomegranate juice. Cook over medium heat until thickened, stirring constantly. Cool, cover and chill.

Pomegranate syrup

Pancakes or waffles are special with this syrup.

Boil together 3½ cups pomegranate juice, 1 teaspoon lemon juice, ⅛ teaspoon salt and ½ bottle liquid pectin or 1 packet pectin. When mixture reaches a boil that cannot be stirred down, add 5½ cups sugar and boil for 5 or 6 minutes. Makes about 3 cups.

Pomegranate pudding

Doris Schultz *of Green Valley also uses pomegranates in this pretty pudding.*

2 cups pomegranate juice
¼ cup sugar
¼ cup cornstarch
¼ teaspoon salt
Small cinnamon stick

To prepare juice, whirl seeds in blender or food processor and strain through a cloth.

Bring juice, sugar, cornstarch, salt and cinnamon stick to a boil. Cook over moderate heat, stirring constantly until mixture thickens. Remove cinnamon stick. Pour thickened juice into pudding dish or individual dishes. Cover with plastic wrap and chill. Serve with cream or custard sauce. Serves 4.

Pomegranate ice

*Pink and pretty is a good description for this recipe from Tucsonan **Hanna Lundberg**.*

1 tablespoon unflavored gelatin
¼ cup cold water
¾ cup boiling water
¾ cup sugar
2 cups pomegranate juice
4 tablespoons lemon or sour orange juice

Soften gelatin in cold water. Add boiling water and stir to dissolve. Stir in sugar; add fruit juices. Cool and freeze in refrigerator tray.

Pomegranates

Pomegranate jelly I

*Pomegranates make outstanding jelly, and this recipe from **Dotty M. Green** of Willcox is a never-fail formula.*

3½ cups pomegranate juice
½ cup lemon juice
7½ cups (3¾ pounds) sugar
½ bottle liquid pectin

To prepare juice, separate and crush the edible portions of 10 to 12 fully ripe pomegranates. Do not remove the seeds. Place fruit in dampened jelly cloth or bag and squeeze out juice. Small amount of water may be added, if needed, to obtain the required amount of juice. Measure sugar and juice into a large saucepan and mix. Bring to a boil over hottest fire, and at once add pectin, stirring constantly. Bring to a full rolling boil (one that cannot be stirred down) and boil for ½ minutes. Remove from fire, skim and pour quickly into glasses. Add paraffin. Makes 11 6-ounce glasses.

Note: If desired, pomegranates may be cut in half and the juice extracted on an orange-juice squeezer. Jelly made with this juice will not be as clear as that put through a jelly bag.

Pomegranate jelly II

Lela Barkley *of Phoenix makes her pomegranate jelly without adding citrus juice.*

Cut in half and juice as you would an orange enough pomegranates to make 4 cups of juice. Strain the juice through a wire sieve and then strain through 4 thicknesses of cheese cloth.

To 4 cups juice, add 7 cups sugar and 1 bottle liquid pectin. Cook until the mixture will jell after a small quantity is dropped into cold water. Pour into jelly glasses, cover with paraffin and seal.

Etiquette note

What do you do with pomegranate seeds when they are served as topping on fruit cocktail or salad?

There appear to be two schools of thought as to how to handle them. In both, you chew the bright pulp from around the seeds, as you would an olive.

Then you have a choice: Take them out of the mouth discreetly and place them on the side of the plate. Or, do as some brave souls do: Swallow them. They are a bit larger than grape seeds.

Grenadine

This is the lovely red syrup that tints and flavors drinks. To make it, combine equal parts of pomegranate juice and sugar. Simmer until thickened and store covered. If refrigerated, it may need to be brought to room temperature before using.

Tequila sunrise

Use an 8-ounce glass to make this layered drink. First put in the glass 3 good dashes of grenadine. Fill with ice cubes and add 1½ ounces tequila. Fill the glass with orange juice. Do not stir.

Quinces

Botanical name: *Cydonia oblonga.*

This fruit is not well known today in home orchards, though once it was regarded as a health food and was widely admired in western Asia and Mediterranean countries. In ancient Greece, quinces were exchanged as tokens of love, despite their very acidic taste.

Less hardy than pear trees, quince trees grow in much the same soil conditions. They do not like cold weather.

The quince is a light, orange-colored fall fruit that is fragrant and similar to a pear in shape. Its nutrients also resemble those of the pear, with about half the calories. It is generally cooked into jellies, preserves and marmalades. (The Portuguese name for quince is *marmelo.*) Because of its high pectin content, it is ideal for combining in jellies with berries or grapes that are low in pectin.

The fruit of a close relative, the flowering quince, also can be used for preserves.

Quinces grow best in elevations of 1,200 to 3,500 feet. Pineapple is the variety generally recommended.

Empanadas de membrillo (Quince pastries)

Rosario Maycher, *Nogales, finds these rich miniature fruit pies a delicious choice for dessert.*

Filling:
  8 ripe quinces
  ¾ cup water
  Pinch salt
1½ cups sugar

Wash, peel and core the quinces; cut into small pieces, add water and salt and cook until tender (about 15 minutes). Mash and add sugar. Continue cooking and stirring for about 15 minutes. Use to fill pastries.

Pastry: ½ cup margarine or butter

1 large egg at room temperature
1¼ cups flour
1 tablespoon water

Whip butter and add egg, using a fork to stir. Add to the flour, cutting into pieces the size of small pebbles. Roll dough to make a large circle and cut with 3-inch round cookie cutter. Put 1 tablespooon quince filling on each circle, fold in half and press the edges together with a fork. Bake at 350 degrees on cookie sheet for about 15 minutes. Makes 12.

Baked quinces

This recipe, taken from an old cookbook, is an interesting dessert suggestion.

    4 to 6 ripe quinces
    6 orange slices
    ¾ cup granulated sugar
    1½ cups water

Wash and dry quinces. Cut, core, pare and slice, and arrange in layers with orange slices in casserole. Sprinkle with sugar and add water. Cover and bake at 300 degrees until tender (1½ to 2 hours). Serves 6.

Quince preserves

Combined with lemons and oranges, quinces make very nice preserves.

    ½ pound prepared quinces
    ½ pound sugar
      Water
      Lemon and orange slices, as required

Pare quinces and cut them in eighths. Reserve the peeling. Remove and discard the cores. Cover peelings with water, measuring the amount needed. For each pint of water, use ½ a sliced lemon (seeds removed) and ½ a sliced orange (seeds removed). Cook quince peelings with water, lemon and orange slices until tender. Strain the juice. Add quince slices to strained juice and cook until fruit is almost tender. Add sugar and continue cooking until fruit is tender. Place fruit in jars. Boil syrup until thick and pour over quince slices. Seal.

Oldfather quince preserves

*When she was a girl, **Beth Oldfather Gladden** of Marana first ate quince preserves made from fruit grown on the family farm in St. David.*

Prepare quinces by peeling, quartering and coring. Place in large kettle and add 3 cups water to each cup of prepared fruit. Simmer for 30 to 45 minutes until fruit is soft. Remeasure and add 1 cup sugar to each cup fruit. Cook again, stirring often with a wooden spoon, until mixture is very thick, clear and honey colored. Pour into hot sterilized jars and seal. Half pears or apples may be used instead of all quinces.

Saguaros

Nuts, seeds and desert beans

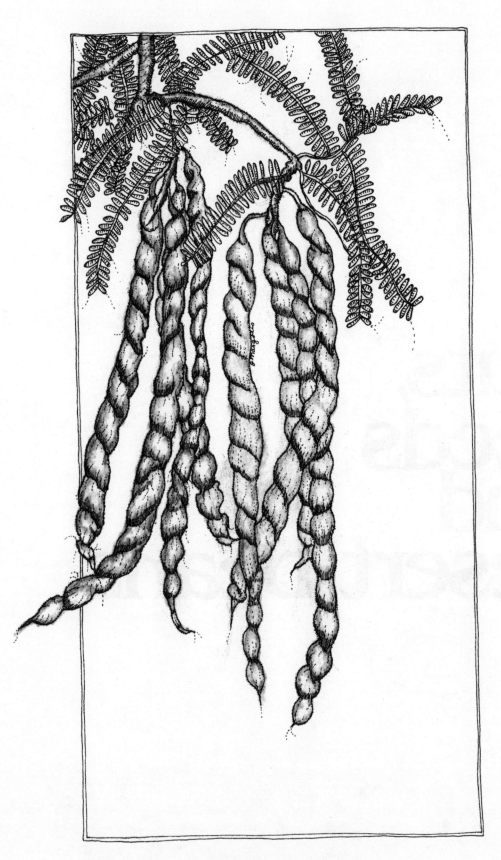

Mesquite beans are intriguing for their sweet flavor.

Nuts, seeds and desert beans

Some home gardeners in the Sonoran Desert and adjoining areas are happily raising nuts, especially pecans and almonds. Pistachios are relative newcomers, but they, too, can succeed in the right environment.

When available, wild nuts can add enjoyment to eating. Like the cultivated ones, they are outstanding sources of protein, oil and B vitamins.

It used to be an annual event to gather acorns, piñons and walnuts in season, and where stands of wild nut trees remain, hikers can still enjoy this custom. The wild ones, too, are sometimes grown in urban yards.

Like nuts, seeds are natural storehouses of nutrients, providing vital vitamins and minerals in important concentrations. They also are a good source of protein and oil.

The bean pods of some plants are sometimes used with or instead of the seeds themselves, especially when the seeds are hard. The pods furnish carbohydrates and B vitamins, but their oil and protein content is minimal.

Acorns

Botanical name: *Quercus emoryi; Q. arizonica*

The best acorns among Southwestern oaks come from the Emory oak or *bellota*. They are tiny but tastier than acorns from other oaks, with a flavor often described as "like an almond, only more bitter."

The shells can be cracked between the teeth. But each acorn should be inspected carefully before cracking, to avoid those with insects.

Today's Native American women shell and grind the acorns to use as they would grated cheese — sprinkled on top of pizzas or vegetables. In the past, they used the acorns as a high-protein staple that could be eaten by itself or used to enrich bread. The nuts give a delicious flavor to modern cookies.

In early summer, acorns are gathered in Mexico, where the season is a little earlier than in Arizona, for sale at stores that specialize in Mexican foods. The tree is common in foothills and canyons of 3,000 to 8,000 feet. About midsummer, the nuts are ready to gather.

Emory oaks can grow as high as 60 feet, and are handsome for landscape use, though slow-growing.

Sweet burritos with Arizona acorns

Acorns are the preferred nuts for these burritos from **Amalia Ruiz Clark** *of Oracle. But if not available, pecans, walnuts or cashews may be subsituted.*

2 small flour tortillas
½ stick butter (melted)
⅓ cup brown sugar
⅓ cup shelled ground acorns
½ cup oil

Brush tortillas with melted butter. Sprinkle with brown sugar. Top with ground acorns. Roll tortillas tightly, fold ends and fasten with toothpicks. Deep-fry burritos until golden brown, place on paper towel to drain. Eat while warm. May be cut in 2 pieces to serve 4.

Almonds

Botanical name: *Prunus amygdalus*

Like its kinsman, the peach, the almond belongs to the rose family. Almond fruit isn't much of anything: a thin pulp that surrounds the seed (nut). When the fruit dries, the fruit splits open to expose the nut. Inside is the edible kernel in its familiar, thin brown covering.

This beloved nut is a native of the Middle East, Eastern Mediterranean countries and North Africa, and since biblical times, it has been cultivated abundantly in the Holy Land. The Moors carried it to Spain, and from there Spanish missionaries brought the trees to California in the 18th century, where they found a ready home.

Almonds come sweet and bitter — the sweet for eating, the bitter for oil and flavoring. The bitter almond is often used as rootstock for the sweet almond and other fruits. Delicate pink or white blossoms make the trees a handsome sight.

The almond tree grows to 25 feet or more in height. It does better at elevations of about 3,500 feet, although the trees will grow at lower elevations, especially the self-fertilizing All-in-One variety. Generally, almonds require cross-pollination, such as Ne-Plus Ultra paired with Mission or Nonpareil.

Chicken almond mousse

Tucsonan **Ethel C. Rounds** *enjoys serving this mousse as the main dish for luncheon or supper.*

2 tablespoons plain gelatin
½ cup white wine
3 egg yolks
1 cup milk
1 cup chicken stock
1 cup ground cooked chicken
½ cup ground almonds
2 tablespoons chopped parsley
1 teaspoon lemon juice
1 teaspoon onion juice
   Salt, celery salt, paprika
   Dash cayenne
1 cup whipped cream

Soften gelatin in wine. Beat yolks slightly in top of double boiler. Add milk and chicken stock. Cook over hot water, stirring until mixture coats a metal spoon. Remove from heat and pour into large bowl. Add gelatin and stir until it dissolves. Add the chicken, almonds, parsley, lemon juice, onion juice and seasonings. Chill until mixture begins to thicken. Fold in whipped cream. Turn into 9-inch ring mold and chill until firm. Serves 8.

Hawaiian chicken salad

Betty Accomazzo *of Laveen brightens her chicken salad with fruits and slivered almonds.*

2 large oranges
2 fresh pineapples
2 cups diced cooked chicken
1 cup chopped celery
1 cup slivered almonds
   Special dressing (see below)

Section and dice oranges, reserving 2 tablespoons juice for dressing. Chill. Cut pineapples in half lengthwise, through the leafy tops. Cut a lengthwise slice, 1 inch thick, from top of each half and a crosswise slice, from each half, about ⅓ from the bottom. Pare, core and chunk the pineapple from the 1-inch-thick pieces and the ⅓ pieces, leaving ⅔ portion with the green tops for the salad presentation. You will need 1 cup pineapple chunks for the salad.

Combine diced oranges, pineapple chunks, chicken, celery and almonds. Blend with dressing. To serve, join the ⅔ portions of pineapple end to end, and pile with half the chicken salad. Garnish with orange sections (other fruits, such as strawberries, pink grapefruit sections, etc., may be used). Repeat arrangement with remaining ⅔ portions pineapple and chicken. Serves 6.

Dressing: Blend 1 cup mayonnaise, ¼ teaspoon salt, 1 teaspoon marjoram, 2 tablespoons orange juice.

Almond-crunch asparagus

Almonds turn this vegetable recipe from **Joan Halper** *of Tucson into a super dish.*

2 pounds fresh asparagus, cooked
1 tablespoon parsley, chopped
1 cup coarse bread crumbs
½ cup ground almonds
4 tablespoons butter

Sauté the parsley, crumbs and almonds in butter until golden brown. Place the hot drained asparagus on a heated platter and top with crumb mixture. Serves 4 to 5.

Chinese chicken

Tucsonan **I-Chen Wu's** *chicken dish may be made in a wok or skillet.*

3 tablespoons oil
½ teaspoon salt
1 cup uncooked chicken breast
½ cup water chestnuts, diced
½ cup bamboo shoots, diced
½ cup diced celery
¼ cup button mushrooms, diced
½ cup chicken stock
1 teaspoon soy sauce
2 tablespoons cornstarch
2 tablespoons water
  Almonds

Heat oil and salt in wok or skillet to sizzling. Add chicken cut in 1-inch cubes and toss and turn at high heat until almost done (about 5 minutes). Reduce heat to medium and add water chestnuts, bamboo shoots, celery and mushrooms. Increase heat to high; toss and turn ingredients until all are blended (about 3 minutes).

Add chicken stock and soy sauce. Cover and cook over medium heat for 5 minutes. Uncover; gradually add cornstarch blended with water. Continue to cook at medium-high heat, turning mixture rapidly until liquids thicken and mixture is very hot (about 3 minutes). Garnish with almonds.

Saucy almonds

A great idea for serving at a cocktail party or as a snack, these crunchy almonds are seasoned with soy sauce. Toss 2 cups whole blanched almonds in 2 teaspoons oil to coat. Spread in single layer in shallow baking pan; bake at 300 degrees for 15 to 20 minutes until golden brown. Transfer to small bowl and toss with 2 teaspoons soy sauce. Return to pan and bake for 5 minutes longer. Cool. Makes 2 cups.

Chicken and almond salad

Kathryn Mansker *of Sierra Vista brought home this recipe from Washington state after a visit with her sister several years ago. Fruits and nuts add an interesting touch.*

3 cups diced cooked chicken
2 tablespoons chopped green onion
2 tablespoons chopped capers
1 teaspoon lemon juice
½ teaspoon salt
1 cup diced celery
1 cup diced orange segments
1 large can pineapple chunks, drained
½ cup toasted almond slivers
½ cup mayonnaise
½ teaspoon grated lemon rind

Mix chicken with onion, capers, lemon juice and salt. Cover and chill for several hours. Just before serving, add celery, orange segments, pineapple chunks and almonds. Combine mayonnaise and lemon rind and mix carefully. Serves 6.

Almond torte

For a very special occasion, this dessert cannot be topped, says **Carolyn Watchman** *of Tucson. For Passover, she makes it with matzo meal.*

8 eggs, separated
1½ cups sugar
  Pinch salt
  Grated rind of 1 lemon
2½ cups blanched almonds, grated
¼ cup bread crumbs

Beat yolks and sugar and salt. Add rind and fold in egg whites that have been stiffly beaten, the almonds and the bread crumbs, in that order. Bake in greased and floured 9-inch spring-form cake pan at 350 degrees for 45 minutes to 1 hour. Great with whipped cream or rum sauce.

Baklava

Anne Houpis *of Tucson enjoys making this delicate Greek sweet, but she no longer prepares the pastry as her mother did. Now she buys it at an Italian import grocery store.*

1 pound finely ground blanched almonds or pecans
½ cup sugar
1 teaspoon cinnamon
½ teaspoon allspice
¼ teaspoon nutmeg
1 pound melted sweet butter
1 pound phyllo pastry sheets
Whole cloves
2 cups honey

Combine nuts, sugar and spices except cloves in large bowl. Melt butter. Brush bottom of 9-by-12-inch pan with melted butter; place one pastry sheet in bottom and brush with butter. Repeat until you have 4 sheets of pastry, each brushed with butter. Spread about 1 cup of nut mixture on pastry. Add 4 more pastry sheets, brushing each with butter. Then a layer of nuts. Repeat process, ending with 4 layers of pastry on top, each brushed with butter.

With a sharp knife, cut through the layers to the bottom to form diamond shapes, place one clove bud in center of each diamond. Bake 1 hour at 300 degrees, until golden brown. Immediately upon removing from oven, slowly pour 2 cups room-temperature honey over top. Cool thoroughly. Let sit overnight or 24 hours for flavor to reach its peak.

Almond-fruit pemmican

Julie Kipps and Renee Losee *of Tucson have found this crunchy nut-and-fruit bar great for supplying energy on a backpack trip.*

¾ cup chopped almonds
¾ cup chopped pecans or other nuts
1 cup chopped dates, chopped apricots or raisins
½ cup honey
½ cup non-fat dry milk
½ cup wheat germ
⅓ cup soy flour
¼ cup wheat bran
2 tablespoons corn oil
Grape or apple juice

Mix nuts, fruit, honey, milk, wheat germ, soy flour, wheat bran and corn oil. Mix in enough juice to make a thick batter. Spread mixture in 8-by-8-inch square pan. Bake at 300 degrees for 30 to 40 minutes or until firm. Cut in squares and allow to cool in pan before removing.

Almond crunch

Tucsonan **Carol Noon** *says these cookies are fabulous, and easy to make.*

2 cups almonds
1¼ cups flour
¼ cup sugar
1 cup butter
1 teaspoon vanilla
1 cup powdered sugar

Grind 1⅔ cups almonds fine. Chop the remainder fine and set aside. Mix flour, sugar and ground almonds. With the fingers work in the butter and vanilla until mixture cleans the bowl. Chill for about 1 hour, roll dough into 1¼-inch balls and then shape balls into rolls 3½ inches long. Form into crescents. Press tops into chopped almonds. Bake on ungreased cookie sheets at 350 degrees for 12 to 15 minutes. Cool for 10 minutes on pan, and roll in powdered sugar. Makes 25 to 30.

Granola

This mixture from Tucsonan **Syd Clayton** *is really delicious and oh, so healthy.*

6 cups oats, preferably from
   health-food store
1 cup whole-wheat flour
1 cup wheat germ
1 cup shredded coconut
1 cup chopped dried prunes
1 cup chopped dried apples
1 cup raisins
1 cup dried banana chips
1 cup chopped dried apricots
2 cups chopped almonds
1 cup sunflower seeds
1 cup honey
½ cup brown sugar
1 cup oil
1 cup dry milk
½ cup water

Put all ingredients in a roasting pan and cook in oven over low heat for about 1 hour, stirring often. When mixture cools, store in Mason jars.

Crazy crunch

A corn-syrup label provided **Norma Entrekin** *of Tucson with this delicious snack idea about 15 years ago. It's still great.*

2 quarts popped corn
1⅓ cups sugar
1 cup margarine
⅔ cup almonds
1⅓ cups pecans
½ cup white corn syrup
1 teaspoon vanilla

Mix popcorn and nuts on cookie sheet. Mix sugar, margarine and syrup in a 1½-quart pan. Bring to boil over medium heat, stirring constantly. Continue boiling, stirring occasionally, for 10 to 20 minutes or until mixture is a light caramel color. Remove from heat. Stir in vanilla. Pour over popcorn and nuts and mix to coat. Spread out to dry. Break apart and store in tightly covered container. Makes 2 pounds.

Almendrado

Mary Catherine Ronstadt *of Tucson lists the original El Charro restaurant as the source of her recipe.*

1 tablespoon gelatin
¼ cup cold water
1 cup boiling water
1 cup sugar
5 egg whites
½ teaspoon almond extract
   Red and green food coloring
   Sauce
1 cup chopped almonds

Soak gelatin in cold water; add boiling water and sugar; stir to dissolve. Chill until slightly congealed. Beat until frothy. Beat egg whites until stiff and fold into gelatin mixture. Add flavoring. Divide into thirds, tinting 1 red and 1 green and leaving the third white — the three colors of the Mexican flag. Layer in oblong mold, beginning and ending with red or green. Chill. To serve, unmold and slice. Top with sauce. Sprinkle with almonds.

Sauce: Scald 1 pint milk in double boiler. Beat 5 egg yolks lightly. Add ¼ cup sugar and ⅛ teaspoon salt, and beat until light. Pour milk over eggs, stir and return to double boiler. Cook and stir until mixture coats spoon. Chill and add ½ teaspoon vanilla.

Carobs

Botanical name: *Ceratonia siliqua*

A widespreading evergreen imported from the Eastern Mediterranean, the carob tree supplies welcome shade in the desert, though its beans and flowers tend to litter. It lives well beyond 100 years, prefers the low-altitude desert, but can grow in protected areas of the midaltitude.

For many people, the beauty of the male carob tree is largely offset by the somewhat displeasing odor of its pinkish blossoms. The female tree has small red blossoms that later become chocolate-brown pods up to 12 inches long. These can be processed into a sweet powder for a cocoa substitute containing no caffeine.

At one time numerous carobs grew on the campus of Arizona State University in Tempe. They are gradually being replaced by other trees.

Also known as locust bean trees, carobs like the same growing conditions as ironwood and citrus, and are drought-resistant. If not pruned, the trees branch close to the ground, and are sometimes used as a hedge. If pruned, they grow to about 30- or 40-foot trees.

Besides carbohydrate and B vitamins, carob powder is a fair source of potassium, and is low in sodium.

Preparing carob powder

Two methods for preparing carob pods are outlined in a bulletin prepared by **Priscilla J. Mays** *of Phoenix, Maricopa County Cooperative Extension Service.*

Gather pods in late August to October when they have reached a deep brown color. To check for ripeness, break a bean pod open and taste the pod by chewing a small piece. It should have a slightly chocolate flavor, with some sweetness. Undesirable pods are less than an inch wide, with thin hard walls and no soft pulp inside.

Pressure cooker method:

Wash pods (no need to open and remove the seeds) and put in a pressure cooker with 1 cup water. Do not crowd cooker. In a 2½-quart cooker, ½ pound of pods is sufficient. Cook at 15 pounds pressure for 20 minutes; cool down immediately under running water.

Slit pods, remove and discard seeds. Discard any pods that did not absorb water. To avoid staining the hands, wear gloves.

Break pods into pieces and dry them in a pan overnight in the oven. In a gas oven, the pilot

Using homemade carob powder

The homemade carob powder is more granular than that made commercially, and is somewhat higher in sugar content than cocoa. Therefore, variations of carob and cocoa recipes need to be tried. Some people may not care for beverages made with homemade carob.

For best results, mix the carob powder with dry ingredients, or dissolve it in milk before adding to other ingredients.

light will give enough heat. In an electric oven, heat oven to 150 degrees and turn off before putting pods in the oven. In the morning, repeat process if pods are not brittle.

Place a few pods at a time in a blender. Start at low speed, then increase to high speed, to reduce the pods to a powder. Discard any unground portions. Powder should be somewhat moist and taste sweet. Powder may be roasted again, if desired, at 140 to 150 degrees. Store in airtight container to prevent caking.

Boiling method:

Wash pods and put in kettle with water to cover. Boil until soft and tender, about 1 hour. Slit pods and discard seeds as described above.

Put small amount of pods at a time in a blender, along with some of the cooking liquid, and grind to a mushlike consistency. Spread in a pan in a thin layer (about ¼ inch) and dry until somewhat brittle in a gas or electric oven, as described above. When dry, break up and put into the blender again, to reduce to a powder. Store in airtight container to prevent caking.

Carob brownie mix

These brownies are especially useful for children and others who are allergic to chocolate or do not want the caffeine that comes in chocolate.

4 cups sifted flour
4 teaspoons baking powder
2 teaspoons salt
4 cups sugar
2½ cups homemade carob powder
2 cups shortening

Mix all ingredients except shortening in a large bowl. Cut in shortening with a pastry blender or a fork until mix is like coarse cornmeal. Store mix in a cool place in a container with tightfitting lid. Makes 12 cups mix.

To make brownies:
2 eggs
1 teaspooon vanilla
2 cups brownie mix
⅔ cup chopped nuts

Beat eggs. Add vanilla and brownie mix. Blend. Batter will not be smooth. Add nuts. Spoon mixture into greased 8-inch square pan. Bake at 350 degrees for 20 to 25 minutes. If glass pan is used, bake at 325. Brownies will pull away from pan when done. Makes 16 brownies.

Mesquites

Botanical name: *Prosopsis juliflora;*
P. pubescens (screwbean)

Long a source of food for Southwestern Indians, mesquite trees are widespread in the desert. The leaves are fernlike and the trees are often used for shade in urban yards, even though the fruit (beans) litter badly when they fall.

The blossoms of mesquite trees, called catkins, furnish pollen for a delicious honey. The pods grow to 6 or 8 inches long and do not split open when ripe, but enclose very hard seeds surrounded by a yellowish, mealy substance that is agreeable and sweet. The preferred pods are those streaked with pink or purple, or well-mottled with red. Honey mesquite beans are straight or curved; those of the less widespread screwbean mesquite develop a twisted bean pod. The pods ripen from early to late summer, often developing a strong "second crop" in response to July and August rains.

Mesquite beans (pods and seeds) were boiled for a flavorful broth or dried and ground into flour by the Indians. (Portions too hard to grind were removed.) Today, the beans can be more completely ground in a sturdy metal grinder or electric blender. The flour, minus any unground parts, is very good in modern muffins and other baked goods.

Because of their natural sugar, mesquite bean pods can be made into a pudding that requires little if any added sweetening. They also make a flavorful jelly.

Mesquites are controversial in some areas. Ranchers claim the trees use water needed for growing feed for cattle. Mesquite supporters maintain the trees control soil erosion, and supply cover and food for wildlife. Efforts to remove mesquites from the list of state-protected native plants have so far failed.

Turning the beans into flour

Gather mature mesquite beans and dry them thoroughly in a low (140 to 150 degrees) oven. Grind with a sturdy grinder. The pods will grind easily, but the seeds are very hard. Discard unground portions and store flour in a tight container (use the refrigerator if there is room as the flour is very appealing to insects).

This is a hearty, coarse flour that can be used in any baking where whole-grain flour is suitable. The best proportion is ¼ cup mesquite-bean flour to ¾ cup regular flour. Be sure to reduce the amount of sugar, if the recipe calls for it, as mesquite-bean flour is sweet.

Mesquite bean coffeecake

Coffeecake made with mesquite-bean flour proved to be quite tasty, **Marsha Alterman and Christine L. Winters** *found in work they did in a University of Arizona class in experimental foods.*

1 cup all-purpose flour
¼ cup mesquite-bean flour
¼ teaspoon salt
½ cup light brown sugar
⅓ cup safflower oil
1 teaspoon baking powder
¼ teaspoon soda
¼ teaspoon cinnamon
⅛ teaspoon nutmeg
½ cup milk
1 well-beaten egg

Combine white flour, salt, sugar and 4 tablespoons oil. Mix until crumbly. Reserve ¼ cup. To the remaining flour mixture add mesquite flour, baking powder, remaining oil, soda and spices. Mix thoroughly. Add milk and egg. Mix well. Pour into greased 8-inch square pan. Spread with reserved ¼ cup crumbly mixture. Bake at 375 degrees for 25 minutes. Makes 1 cake.

Mesquite-bean popovers

Carlos Nagel of Tucson uses mesquite-bean flour to make various kinds of bread, including popovers.

¼ cup mesquite-bean flour
¾ cup all-purpose unbleached flour
¼ teaspoon salt
2 large eggs
1 cup milk
1 tablespoon melted butter or shortening

Grease popover pans or large muffin pans well and heat for 10 minutes in 450-degree oven. Combine flours and salt. Beat eggs in separate bowl until frothy. Add milk and melted butter. Stir liquid ingredients slowly into dry ingredients and beat just until well-blended. Pour batter into heated pans, filling about ½ full. Bake on center rack at 400 degrees for about 40 minutes. Makes 8 to 10 popovers.

Atole

Herlinda Molina of Tucson passes along a friend's directions for preparing atole, a non-alcoholic drink made with mesquite beans and once widely used by the Indians of the Sonoran Desert.

Gather mesquite beans (preferably those with red or purple on the pods) and cover with water. Boil for up to an hour to soften. Cool and mash, strain the pulp to get the seeds out. Heat the liquid, adding cinnamon stick and sweetening to taste with white sugar, brown sugar or Mexican raw sugar (panocha). Thicken with a little flour or cornstarch.

Black bottom mesquite cupcakes

When she can get mesquite flour that is ground fairly fine, Tucsonan Stephanie Daniel likes to make these fancy cupcakes.

1 8-ounce package cream cheese
1 egg
⅓ cup sugar
⅛ teaspoon salt
1 cup semi-chocolate chips
1⅓ cups fine-ground mesquite flour
¼ cup cocoa
1 teaspoon baking soda
½ teaspoon salt
1 cup water
⅓ cup oil
1 tablespoon white vinegar
1 teaspoon vanilla
1 cup sweetened whipped cream

Using wooden spoon, blend softened cream cheese with egg, sugar and salt in mixing bowl. Carefully fold in chocolate chips and set aside.

Combine mesquite flour, cocoa, soda and salt and mix well. Add water, oil, vinegar and vanilla and blend thoroughly. Fill 12 paper-lined muffin tins ¾ full and drop 1 tablespoon cream cheese mixture in center of each. Bake for 35 to 40 minutes at 375 degrees. Spoon 1 tablespoon sweetened whipped cream onto the center of each cooled cupcake. Makes 12.

Mesquite bean butter

Interest in desert food sources led **Lance Grebner**, *a University of Arizona graduate student from Illinois, to develop this recipe.*

  3 quarts ripe mesquite beans
    Water
  4 cups sugar
  ⅓ cup lemon juice
  1 bottle liquid pectin
  1 tablespoon cinnamon

Cut each bean into 2 or 3 pieces and cook in vegetable steamer over boiling water until tender (about 30 minutes), or until mesquite-bean pods pull apart easily. Put cooked pods in blender, 1 cup at a time with ½ cup water, and chop. Then put through coarse strainer. Discard fiber and seeds.

Add water as needed to make 8 cups mesquite pulp. Place in large kettle or saucepan over high heat, stir in sugar and lemon juice, and bring to a full, rolling boil. Boil hard for 1 minute. Remove from heat, stir in pectin and bring to a boil again.

"Butter" is ready when a small amount dropped in a bowl of water forms a soft ball. Finally, add cinnamon, stir and pour into hot, sterilized jars and seal with paraffin.

Mesquite bean jelly

The sweetest and best beans (fruit) of the mesquite are those that are streaked or well-mottled with pink or red, say the experts.

  3 quarts mesquite beans
    Water
  4½ cups sugar
  4 tablespoons lemon juice
  1 package powdered pectin

Cut each mesquite bean into 2 or 3 pieces, place in large kettle and add water to cover. Simmer until liquid turns yellow. Strain. You will need 3 cups of liquid. Place the liquid in a kettle or large saucepan, stir in powdered pectin and cook. Stir constantly over high heat until mixture comes to a boil. Add sugar and lemon juice and stir. Bring to a full, rolling boil, stirring constantly. Boil hard for 1 minute, or until syrup comes off metal spoon in a sheet. Remove from heat, skim off foam with metal spoon and quickly pour into sterilized glasses. Cover at once with hot paraffin.

Note: If desired, red food coloring may be added.

Palo verdes

Botanical names: *Cercidium floridum* (blue palo verde) and C. microphyllum (foothill palo verde)

Two palo verdes, the blue and foothill varieties, jointly carry the title of Arizona state tree. Each produces a bright halo of yellow blossoms in spring that add enchantment to the desert. In a culinary book, both are of interest because of their edible bean pods.

The blue palo verde grows naturally along washes to about 25 feet in height. Its bark and foliage are blue-green. Its beans or fruit are about 3 inches long. When tender and green (they somewhat resemble green beans), they can be steamed for a tasty green vegetable that furnishes a fair amount of vitamin A.

Or they can be shelled and the seeds cooked like garden peas or used raw in salads. After the beans mature, the seeds can be removed from the pods, parched and ground into flour.

The foothill palo verde (also known as little-leaf or mesa palo verde) is a little smaller and blooms somewhat later than the blue. Its bark is yellow-green. The foothill palo verde is sometimes called the "saguaro nursemaid" because it furnishes shelter to the slow-growing saguaro cactus. Its small beans can be used as described above. Opinions differ as to whether foothill or blue palo verde beans are preferable.

There should be no confusion of the blue and foothill palo verdes with the Mexican palo verde, also found in the desert and used in urban landscapes. Its long, stringy leaves consist of 8- to 14-inch midribs that bear many leaflets an eighth of an inch long. When shed, the midribs create a straw-colored thatch on the ground. The 6-inch pods that follow the spring display of Mexican palo verde's yellow flowers are toxic.

Peanuts

Botanical name: *Arachis hypogaea*

Though treated like nuts in cooking, peanuts belong to the legume (pea and bean) family. Also called goobers or groundnuts, peanuts were grown in South America at least 2,000 years ago, and are now cultivated in many parts of the world. Until the Civil War, they were grown mostly in Virginia in this country.

The plant produces yellow flowers. After pollination, the flowers fade and long shoots grow into the ground. The nuts form at the end of the shoots. Highly nutritious, peanuts are 40 percent to 50 percent oil and 20 percent protein.

The peanut is an annual that can be grown below the 3,500-foot level. Seeds are grown commercially in Yuma County. Individuals, however, can grow the nuts if they have the right sandy soil and use a fair amount of water. A good-producing plant can furnish a half-gallon or more of peanuts. The Spanish-style peanut is the popular one for growing at home.

Roasting peanuts

Delia Turner, *who raises peanuts at her Patagonia home, tells how to roast them.*

Pull peanuts from the ground when mature (about mid-October), shake well and spread out to dry for about 10 days.

Spread peanuts in shallow pan in cold oven and set thermometer at 350 degrees. Leave heat on for 15 minutes, turn oven off and allow peanuts to cool enough to handle. Then crack and eat 'em.

Peanut butter

Place 1½ cups roasted peanuts, salted or unsalted, in blender; cover and blend to smooth or to nutty peanut-butter consistency. Stop blender from time to time and use rubber spatula to keep mixture around the processing blades. Store in refrigerator. Makes ¾ cup.

Peanut rice loaf

A vegetarian-style main dish featuring peanuts and rice comes from Tucsonan **Nettie Cluff***.*

1 large onion, chopped
1 tablespoon oil
⅓ cup peanut butter
3 cups cooked rice
1½ cups chopped roast peanuts
1 pound cottage cheese
1 cup chopped celery
1 cup diced cooked carrots
4 eggs, beaten
1 teaspoon salt (optional)

Sauté onion in 1 tablespoon oil. Combine with other ingredients and pack into greased loaf pan. Bake at 375 dgrees for 1¼ hours.

Peanut coleslaw

Salted peanuts are the special ingredient that makes this simple coleslaw recipe from **Hazel Coatsworth** *of Tucson special.*

3 cups shredded cabbage
½ cup salted peanuts
⅓ cup salad dressing (Miracle Whip or similar)
2 tablespoons half-and-half
¼ teaspoon salt
2 tablespoons sugar

Lightly combine the cabbage and peanuts. Mix the salad dressing, half-and- half, salt and sugar, and add to the salad. Serves 4.

Peanut brittle

Joy Wolf *of Rio Rico uses excellent peanuts grown near Patagonia to make this recipe, which she says was originated a long time ago by* **Blanche Gardner** *of Nogales.*

1¼ cups sugar
½ cup white corn syrup
½ cup water
Pinch salt
½ pound raw peanuts
3 teaspoons butter
1¾ teaspoons baking soda

Combine sugar, syrup, water and salt. Cook on high heat in cast-iron skillet, stirring with a wooden spoon until the syrup spins a 6-inch thread. Add peanuts and keep cooking until syrup turns light brown. Add butter and soda and stir until butter melts. Pour on greased platter or pizza pan.

Pecans

Botanical name: *Carya illinoensis*

For those interested in shade trees that produce nuts as well, the pecan is a great choice for the home orchard.

This native American nut is more than satisfactory in middle and high desert areas, if it is given deep soil and sufficient irrigation. Commercial production in the Tucson area and in Cochise County is a satisfactory business.

Some growers are successfully reducing water use by means of drip irrigation. Two trees are needed for pollination and better crop yield. The Western Schley is considered best for desert climates.

The rich nuts are ready to gather in late summer or early fall.

Downfall appetizer spread

Barbara Uhrig of Tucson finds this sour cream and pecan appetizer impossible to top. The recipe is from her sister-in-law, **Ruth.**

1 8-ounce package cream cheese
½ cup chopped pecans
½ cup sour cream
1 2½-ounce jar dried beef, finely chopped
2 tablespoons instant minced onion
2 tablespoons finely chopped green pepper
2 tablespoons milk
⅛ teaspoon white pepper

Combine all ingredients and put in 8-inch pie plate or 1-quart casserole. Bake at 350 degrees for 20 minutes or until edges are bubbly. Serve hot as a spread for crackers, toast or melba toast.

Pecan roast

While her husband looks after the family pecan orchard near Benson, **Eleanor Mattausch** *develops good ways to use the nuts, like this well-seasoned main dish.*

1 cup pecans
1 cup ready-to-eat cereal
½ cup chopped onions
2 eggs
½ teaspoon celery salt
1 teaspoon sage
1 tablespoon paprika
½ cup milk

Combine ingredients; bake in greased loaf pan at 350 degrees for 35 minutes.

Pecans

Pecan rice loaf

This pecan loaf from **Brenda O'Berry** *of Thatcher is one her mother used to make.*

1 cup bread crumbs
2 cups chopped pecans
⅔ cup cooked rice
1 medium onion, chopped
1½ cups milk
3 eggs, beaten
1 teaspoon sage
3 tablespoons cooking oil
2 tablespoons chopped parsley
2 tablespoons soy sauce
½ teaspoon salt
Pinch of thyme and sweet basil

Combine ingredients thoroughly. Pour into greased loaf pan and bake for 1 hour at 350 degrees. Serve with a gravy.

Pumpkin-pecan biscuits

*This suggestion from Tucsonan **Sue Scheff** is good any time hot biscuits are desired, but especially during the Thanksgiving to New Year's holiday season.*

2 cups flour
2 to 3 tablespoons sugar
4 teaspoons baking powder
½ teaspoon salt
½ teaspoon cinnamon
½ cup margarine
½ cup chopped pecans
½ cup milk
⅔ cup cooked mashed pumpkin

Stir the dry ingredients together. Cut in margarine to make a crumbly mixture, and stir in nuts. Combine milk and pumpkin. Stir both mixtures together until all is moistened. Dough will be fairly stiff. Turn onto floured board and knead a few times. Roll out ½-inch thick and cut into 2-inch rounds. Place 1 inch apart on greased baking sheets. Bake for 20 minutes at 425 degrees. Serve hot.

Pecan-oatmeal muffins

*Tucsonan **Helen Underwood** recommends these muffins highly. With pecans added, they are a treat for Sunday breakfast. Serve with favorite preserves (she likes peach).*

1 cup quick-cooking oats
1 cup sour milk
1 egg
½ cup brown sugar
¼ cup melted shortening
1 cup flour
½ teaspoon salt
1 teaspoon baking powder
½ teaspoon soda
½ cup chopped pecans

Soak oatmeal in sour milk for 1 hour, then add egg and beat well. Add sugar and mix. Add cooled shortening. Sift together flour, salt, baking powder and soda. Add to moist mixture and stir just until all ingredients are moist. Fold in pecans. Pour into a muffin tin and bake at 400 degrees for 15 to 20 minutes. Makes 1 dozen.

Festive pecan and pimento salad

*This attractive gelatin salad from **June C. Gibbs** of Tucson is not only superb with turkey or chicken, but is great for entertaining because you can get it ready early.*

1 package lemon-flavor gelatin
¼ teaspoon salt
½ teaspoon grated lemon rind
2 tablespoons lemon juice
¼ cup chopped pimento
½ cup chopped pecans
½ cup finely chopped celery
  Crisp lettuce or watercress

Prepare gelatin according to package directions. Add salt, lemon rind and juice. Cool until mixture begins to thicken. Add pimento, pecans and celery. Pour into bell-shaped molds. Chill until firm. Serve on lettuce or with watercress. Recipe fills 6 ½-cup molds.

Savory potato-pecan bake

Rita Rosenberg *of Tucson, experienced in European cooking, was enthusiastic about trying new ingredients when she moved here. This recipe, converted from walnuts to pecans, demonstrates the superb taste and versatility of the American nut.*

2 tablespoons vegetable oil
3 tablespoons chopped onion
2 tablespoons unbleached flour
1 cup half-and-half
4 medium potatoes, boiled and riced
  Salt and pepper
½ teaspoon marjoram
¼ teaspoon nutmeg
1½ cups chopped pecans
  Parsley

In skillet, sauté onions in oil until golden. Add flour, stir and mix. Add small amount of half-and-half; keep stirring while adding all of half-and-half. Stir until smooth and bring to a boil. Taste for seasoning and add rest of ingredients, except parsley. Put in buttered ovenproof dish and bake at 325 degrees for 30 minutes. Sprinkle with parsley. Serves 6 to 8.

Pecan-broccoli company casserole

Rae Eichinger *of Tucson likes to serve this vegetable-nut combination when she entertains. The recipe comes to her from her sister in Oklahoma, where there are a lot of pecan trees, too.*

2 10-ounce packages frozen chopped
  broccoli
1 10¾-ounce can condensed
  cream of mushroom soup
½ cup mayonnaise
¾ cup chopped pecans
1 medium onion, finely chopped (or 1½
  tablespoons dry onion flakes)
2 eggs, well beaten
1 cup grated sharp cheese
2 cups bread crumbs

Cook broccoli according to package directions. Drain. Add soup, mayonnaise and chopped pecans. Mix well. Add onion and eggs. Pour into greased 1½- or 2-quart casserole dish. Sprinkle with grated cheese and top with bread crumbs. Bake at 350 degrees for 30 minutes. Serves 6 to 8.

Pecan-apple stuffed squash

Chopped pecans and apples make acorn squash into a special dish in this recipe from Tucsonan **Mary C. Jones.**

2 acorn squash
½ teaspoon salt
2 medium apples, chopped
1 teaspoon lemon juice
¼ cup chopped pecans
2 teaspoons brown sugar

Cut squash in halves lengthwise and scrape out seeds and fiber. Place halves in baking dish with cut sides down and pour ½-inch water into pan around them. Bake at 400 degrees for 20 minutes or until just tender. Pour off water. Turn squash halves over and sprinkle with salt, then fill with a mixture of apples, nuts and lemon juice. Sprinkle brown sugar over top. Return to oven and bake for 10 minutes or until filling is hot. Serves 4.

Pecan cake with pineapple

This pretty and delicious dessert gets "rave notices" every time **Mary Metzger** *of Casa Grande serves it. It is easily made.*

1½ cups sugar
2 cups flour
2 teaspoons baking soda
½ teaspoon salt
3 eggs
2 teaspoons vanilla
2 tablespoons oil
1 No. 2 can crushed pineapple
½ cup pecans

Mix ingredients for cake and bake in greased and floured 9-by-13-by-2-inch pan at 350 degrees for 35 minutes, or until done. Frost with cream cheese frosting.

Cream cheese frosting: Combine 8 ounces softened cream cheese, 1 tablespoon softened butter or margarine, 2 cups powdered sugar, 2 teaspoons vanilla and 1 cup chopped pecans.

Sad cake

The directions for this cake were obtained by Globe resident **Lenora Clark** *at a family reunion in Texas. She says it's called "sad" because it always falls while baking, but tastes delicious.*

2 cups biscuit mix
2⅓ cups brown sugar
4 eggs
2 cups pecans, chopped
1 teaspoon vanilla

Combine biscuit mix and sugar. Add eggs one at a time, beating well after each egg is added. Fold in chopped pecans and vanilla. Pour into greased 9-by-13-inch baking pan. Bake at 350 degrees for 35 to 40 minutes.

Pecan bread with pumpkin

Daphine G. Gawin *of Casa Grande calls this bread "the working girl's delight" because the recipe makes four loaves. It's great for freezing.*

1 large can pumpkin (or 3 cups mashed)
1 cup oil
1 teaspoon salt
1 teaspoon soda
4 cups sugar
1½ teaspoon cinnamon
1 teaspoon pumpkin pie spice
5 cups flour
1 cup chopped pecans
1 cup raisins (optional)

Mix ingredients in large bowl, except for flour, nuts and raisins, beating until smooth. Add flour gradually and beat again; add nuts. Add raisins, if using. Grease 4 1-pound coffee cans and fill ½ full. Bake at 350 degrees for 1½ hours. Cool in cans. Remove and wrap in foil. Freeze, if desired.

Best pecan pie

When he cooked for Tucson Fire Station No. 1, **Glenn N. Carlton** *of Tucson made this pie often. Now that he's a captain, he doesn't do the cooking, but still makes the pie for special occasions.*

3 eggs
½ cup sugar
Pinch salt
½ teaspoon vanilla
1 cup light corn syrup
1 cup fresh pecans, broken
1 unbaked pie shell

Beat eggs slightly and stir in sugar, salt, vanilla and syrup. Mix well but do not beat. Add pecans, folding in to cover them with syrup. Pour in unbaked pie shell. Bake for 10 minutes at 425 degrees, then for 30 to 35 minutes at 350 degrees. Pie will puff way up, but settles down as it cooks.

Party millionaire pie

Whenever she serves her version of the famous Furr's Cafeteria pie, guests ask **Thelma Romney** *of Green Valley for the recipe.*

3 egg whites
1 cup granulated sugar
½ teaspoon salt
1 teaspoon vanilla
21 Ritz crackers
⅔ cup chopped pecans
1 8-ounce package cream cheese
2 cups powdered sugar
1 cup whipping cream
1 small can crushed pineapple, well-drained
½ cup chopped pecans

Beat egg whites until stiff, adding granulated sugar gradually, along with salt and vanilla. Crush crackers and fold into egg white mixture. Add ⅔ cup pecans. Shape into a 10-inch glass pie plate. Bake at 350 degrees for 30 minutes. Beat cream cheese and powdered sugar until creamy. Fold in the cream, whipped until stiff. Fold in well-drained pineapple and remaining chopped pecans. Spread filling on cooled crust. Refrigerate pie until serving time. Serves 8 to 10.

Pecan-crunch pie shell

Tucsonan **Betty Birkett** *shares this delicious pie-shell recipe, to be filled with lemon or cream pudding.*

Combine ½ cup soft butter with ¼ cup brown sugar, 1 cup sifted flour and ½ cup chopped pecans. Reserve ¾ cup for topping and press remainder into a 13-by-9½-by-2-inch pan. Bake at 400 degrees for 15 minutes. Cool, add favorite lemon or cream pudding, top with balance of nut mixture and chill.

Pecan paradise pie

Grace Pawson *of Tucson sends her easy-to-make pecan pie that was given to her by a neighbor when she lived in Ohio. It was popular for serving at country church dinners at threshing time.*

1 cup granulated sugar
1 cup flour
1 egg, well beaten
¾ teaspoon baking soda
¼ teaspoon salt
3 tablespoons lemon juice
1 No. 2 can fruit cocktail, well drained
½ cup brown sugar
¾ cup chopped pecans

Mix granulated sugar, flour, egg, baking soda, salt and lemon juice. Add drained fruit cocktail, mix well and spread in ungreased pie pan. Mix brown sugar and pecans and use as topping. Bake at 300 degrees for 45 minutes. Can be served with vanilla ice cream or whipped topping.

Pecan pie

Donna Coca *of Morenci finds she must bake this pie at Christmas for her dad.*

3 tablespoons butter
⅔ cup light brown sugar
 Pinch salt
3 eggs
¾ teaspoon vanilla
⅔ cup light corn syrup
½ cup half and half or evaporated milk
1 cup chopped pecans

Cream butter. Slowly beat in sugar and salt. Add eggs one at a time and beat briskly. Blend in remaining ingredients. Pour into unbaked pie shell. Bake for 10 minutes at 450 degrees. Reduce heat to 350 and bake until custard sets, about 30 minutes, or longer if more than one pie is cooked at once. Makes 1 9-inch pie.

Capirotada

A more elaborate version of the bread pudding comes from **Lupe de Santiago** *of Casa Grande.*

11 slices bread
2 sticks cinnamon
10 cloves
2 cups water
1 cup brown sugar
3 small bananas
2 medium apples
1 cup crushed pineapple
1 cup raisins
½ cup chopped pecans or peanuts
½ cup grated Longhorn cheese

Toast bread and break in small pieces. Boil cinnamon and cloves in water briefly and strain. Stir in brown sugar. Peel and chop fruit. To be certain to mix pudding thoroughly, assemble in layers and then mix lightly, this way: Spread bread in bottom of 9-by-12-inch casserole; add layers of fruit, nuts and cheese; repeat. After mixing, pour the sugar-spice liquid over all. Bake at 350 degrees for 30 minutes.

Bea's oatmeal-pecan cookies

Katie Kelly *of Tucson finds cookies made from a friend's recipe always welcomed by her family.*

1½ cups flour
1 teaspoon salt
1 teaspoon soda
1 cup shortening
⅔ cup brown sugar
⅔ cup white sugar
2 unbeaten eggs
1 teaspoon hot water
1 teaspoon vanilla
1 12-ounce package chocolate chips
2 cups oatmeal
1 cup chopped pecans or other nuts

Sift flour, salt and soda; cream sugars and shortening and beat in eggs, then hot water and vanilla. Fold in dry ingredients, chips, oatmeal and pecans. Drop by teaspoonfuls on greased cookie sheets. Bake at 350 degrees until lightly browned. Makes 5 to 6 dozen medium cookies.

Pecan logs

Bev Bergsma *of Bowie, who has a pecan orchard, always has plenty to call on for personal use. Among her cookies is this easy-to-do favorite.*

1⅓ cups flour
¾ cup softened butter
1½ cups chopped pecans
3 tablespoons sugar
¼ teaspoon salt
1½ teaspoons vanilla
¾ cup powdered sugar
¼ cup unsweetened cocoa

In large bowl, combine all ingredients except powdered sugar and cocoa. With hand, mix until blended. Refrigerate for 30 minutes in plastic bag.

Combine ¼ cup powdered sugar and the cocoa and reserve. Reserve remaining sugar separately. Roll out dough, half at a time, to ¼-inch thick. Cut into 3-by-½-inch strips. Place 1 inch apart on ungreased cookie sheet. Bake at 350 degrees for 8 to 10 minutes. Let stand 1 minute. Remove to wire rack. Roll half in powdered sugar and half in cocoa-sugar mixture, while still warm. When cool, roll again, if desired. Makes 5 dozen.

Green Valley pecan crescents

This cookie recipe, sent to us by **Val Johnson** *of Tucson, is one she has used often since it was given to her 20 years ago when traveling in the military with her husband.*

1 cup butter
¾ cup confectioners' sugar
2 cups pecans, coarsely chopped
1 teaspoon vanilla
2 cups flour
1 tablespoon ice water
⅛ teaspoon salt

Cream butter and sugar; add remaining ingredients. Roll with palms of hands into finger-lengths; shape into crescents. Bake on cookie sheets at 325 degrees for 15 to 20 minutes or until firm but not brown. Immediately roll in additional confectioners' sugar while still hot, then again when cooled. Makes 3 dozen cookies.

Swedish nuts

These nuts go in holiday-wrapped tins to special friends at Christmas time from **Barbara Schuelke,** *Phoenix. They are a wonderful change from salted nuts.*

1 pound shelled pecans
2 egg whites
1 cup sugar
½ cup butter

Toast nuts at 325 degrees on a jelly-roll pan until lightly browned — about 20 minutes. Beat egg whites, adding sugar gradually, until stiff peaks form. Fold toasted nuts into meringue. Melt butter in same pan; spread nut mixture over butter. Bake at 325 degrees for 30 minutes, stirring every 10 minutes. Nuts will be coated with golden meringue and all butter will be absorbed. Cool and break apart any nuts that have stuck together. Store in air-tight jar. Makes about 1¼ pounds.

Pecan leaves

Pecan tassies

Ella Otstot *of Green Valley finds this version of the famous pecan dessert an easy-to-make idea for parties.*

Crust:
1 3-ounce package cream cheese
1 stick butter or margarine
1 cup flour

Filling:
1 egg
¾ cup brown sugar
1 tablespoon melted butter
Dash of salt
⅔ cup coarsely broken pecans

Mix crust ingredients well and roll into 24 small balls. Let stand in refrigerator at least 2 hours before using.

Using muffin pan with 12 small cups, press a ball of dough into each to receive the filling. Combine filling ingredients and fill cups half full. Bake at 325 degrees for about 30 minutes or until filling sets. Makes 24 tassies.

Texas carrot cookies

Barbara Bishop *of Tucson says Texas Agriculture Commissioner John C. White should be credited with this recipe, which is both nutritious and appealing.*

⅛ teaspoon baking soda
½ cup honey
½ cup butter
1 egg, slightly beaten
1 cup sifted flour
1 teaspoon baking powder
⅛ teaspoon salt
1 cup quick oatmeal, uncooked
½ cup chopped pecans
⅔ cup grated raw carrots
1 teaspoon vanilla
1 teaspoon cinnamon
¼ teaspoon nutmeg

Combine soda and honey; set aside. Cream butter, add egg, then honey and soda mixture. Mix well. Sift flour, baking powder and salt into creamed mixture. Fold oatmeal, pecans and grated carrots into batter. Add vanilla, cinnamon and nutmeg. Drop by teaspoon onto greased cookie sheet. Bake at 250 degrees for 12 minutes or until golden brown. Makes 50.

Pecan dream bars

Doris C. Hill *of Tucson used to work in an office with a lot of "chowhounds" who loved pecan pies. When she couldn't supply enough to satisfy them, she switched to making these bars. They're great for Christmas gifts, too.*

Crust:
½ cup brown sugar
½ cup butter or margarine
1 cup flour

Filling:
1 cup brown sugar
2 tablespoons flour
½ teaspoon baking powder
2 eggs, beaten
1 teaspoon vanilla
1½ cups coarsely chopped pecans

To make crust, mix butter and sugar and blend in flour. Press into a 12-by-8-inch, lightly greased pan, extending crust slightly up sides. Bake for 12 minutes at 350 degrees. Cool slightly.

To make filling, mix sugar, flour and baking powder. Add to eggs. Then stir in vanilla and nuts. Pour into crust. Bake at 350 degrees for 20 minutes. Cut in squares while warm.

Mexican wedding cakes

These delicious traditional cakes are actually small round cookies. Be careful to bake just until lightly browned, cautions **Lea R. Ward** *of Tucson.*

¾ cup butter
4 tablespoons powdered sugar
2 teaspoons vanilla
1 teaspoon cold water
2 cups sifted all-purpose flour
⅛ teaspoon salt
1 cup pecans, finely chopped
Powdered sugar

Cream butter until fluffy. Add sugar, vanilla and cold water. Mix well and stir in flour, salt and nuts. Chill dough thoroughly.

Make into small balls and place on ungreased cookie sheet. Bake at 400 degrees for 8 to 10 minutes, until slightly browned. Remove from baking sheet and roll in powdered sugar while still hot. Place on rack to cool. Roll again in powdered sugar. Makes 4 dozen.

Note: Dough may be shaped into a roll about 1½ inches in diameter, wrapped in waxed paper and chilled. Cut in ¼-inch slices and place on ungreased cookie sheet. Bake at 400 degrees for 6 to 8 minutes or until lightly browned. Roll in sugar.

Pecan-olive cheese ball

For people short on time to spend in the kitchen, this simple but tasty cheese ball is the welcome idea from **Billie Mauntel** *of Tucson.*

8 ounces cream cheese
4 ounces blue cheese
¼ cup butter or margarine
1 small can chopped olives
2 tablespoons chopped green onions
½ cup chopped pecans

Soften cheeses and butter. Blend well. Add chopped onions and olives. Roll ball in chopped pecans. Make ball a day in advance for flavors to blend. Also can be made ahead and frozen.

Fast-cook fondant

This old-fashioned candy recipe is from **Marie Simmons** *of Safford, who has found that it never fails to please.*

1 pound confectioners' sugar, sifted
⅓ cup butter or margarine
½ cup light corn syrup
1 teaspoon vanilla
¼ cup chopped pecans or other nuts
Food coloring, if desired

Combine half of sugar, the butter and the corn syrup in 3-quart heavy saucepan. Cook over low heat, stirring constantly, until mixture comes to a full boil. Quickly stir in remaining sugar and the vanilla and remove from heat. Stir only until mixture holds its shape. Add nuts and pour into buttered 8-by-9-inch square pan. Cool just enough to handle. Knead until smooth. (If candy gets too hard to knead, work with a spoon to soften, and then knead.) If desired, divide into parts and tint with food color while kneading. Makes 1½ pounds.

Arizona pralines

This hand-me-down from a friend is a favorite with **Beth Estes** *of Double Adobe.*

2½ cups granulated sugar
1 teaspoon baking soda
1 cup buttermilk
¼ teaspoon salt
3 tablespoons butter
2⅓ cups pecan halves

Using a large pan (8-quart), mix first 4 ingredients together. Cook on high for 5 minutes, stirring often and scraping bottom of pan. Add butter and nuts and cook for another 5 minutes, stirring constantly. Remove from heat and let cool for 2 minutes. Beat with spoon until thick and creamy. Working fast, drop by tablespoonfuls onto wax paper. Cool.

Holiday pecans

Roslyn B. Sirota *of Green Valley used to enjoy this recipe with walnuts when she was growing up on a Connecticut farm. Now she uses Arizona pecans.*

2 cups pecan halves
1 tablespoon egg white
¼ cup sugar
1 tablespoon cinnamon
⅛ teaspoon cloves
½ teaspoon nutmeg

Place nuts in small bowl and pour egg white over them. Stir until nuts are coated and sticky. Mix sugar and spices and sprinkle over nuts, stirring until sugar mixture completely coats nuts. Spread on ungreased baking sheet and bake at 300 degrees for 30 minutes. Makes 2 cups.

Pecan-cheese ball

*Tucsonan **Cyndi Fraser's** cheese ball always draws compliments. It can be made ahead of time and frozen.*

2 5-ounce jars Old English cheese spread
1 5-ounce jar Roka blue cheese spread
1 8-ounce package cream cheese
1 tablespoon Worcestershire sauce
    Garlic salt to taste
    Chopped pecans

Have cheeses at room temperature and mix all ingredients together. Form into 1 large ball or 2 small ones. Roll in pecans. Serve with crackers.

Pecan caramels

*Tucsonan **Julee McGill** finds these candies are excellent for gifts.*

1 cup white sugar
¾ cup dark corn syrup
½ cup butter
1 cup half-and-half
1 cup broken toasted pecans
½ teaspoon vanilla extract

In heavy pan, combine sugar, syrup, butter and ½ cup cream. Using wooden spoon, bring mixture to rolling boil, stirring constantly. Add remaining ½ cup half-and-half and stir slowly to hard ball stage (260 degrees on candy thermometer). Remove from heat. Add pecans and vanilla. Pour into 8-inch square, buttered pan. When partially cool, mark into squares. Cut when cold. Wrap individually in waxed paper and store in airtight container. Makes 36 pieces.

Elly's caramels

*Homemade candies are the best kind, and this is one that's easy to do from **Rosejean Hinsdale** of Scottsdale.*

2 cups sugar
½ teaspoon salt
1 cup light corn syrup
2 cups warm light cream
⅓ cup butter
1 teaspoon vanilla
½ cup pecans

Mix sugar, salt, corn syrup and 1 cup of the cream a in large saucepan. Cook, stirring, for about 10 minutes. Add remaining cream very slowly so mixture does not stop boiling. Cook for 5 minutes longer. Stir in butter, 1 teaspoon at a time, and cook slowly to 248 degrees. Remove from heat. Add vanilla and pecans. Mix gently. Pour into buttered 8-by-8-by-2-inch pan. Cool. Turn out on board and mark off in ¾-inch squares. Cut and wrap individually in waxed paper.

Sugared nuts

*This is an easy-do idea for serving at the holidays from **Lunette Cole** of Tucson.*

Cook together 1 cup sugar and ⅓ water until syrup spins a thread from a spoon. Remove from heat and stir in 1 heaping tablespoon butter, 1 teaspoon cinnamon and 2 cups pecan halves. Stir until nuts are sugared and separated.

Piñons

Botanical name: *Pinus edulis, P. monophylla*

The small piñon nuts or pine nuts of the Southwest are sweet and delicious. They were long used by the Indians in dozens of ways, including grinding both nut and shell into flour. Piñons are very high in fat content (about 60 percent), contain about 15 percent protein and 17 percent carbohydrate.

The trees also were important to the Indians for another reason: A lush output of spring blossoms was considered an indication that the coming growing season would be a good one for various crops.

Piñons are gathered for sale by the Hopi and Navajo in northern Arizona, but the nuts also can be found in limited amounts in cooler elevations (up to 7,000 feet) in the southern half of the state.

The picturesque piñon has horizontal branches and is smaller than other pines. In the wild, piñons become irregular in shape with age; in urban yards they can be pruned into excellent ornamentals. They do not, however, grow well in low desert settings.

Piñon cakes

Tucsonan **Martha "Muffin" Burgess** *passes along this old Indian recipe that is sweet and flavorful, but extremely rich.*

2 cups shelled piñon nuts
¾ cups water
½ teaspoon salt
2 tablespoons oil

Purée nuts into meal and mix with water and salt. Let stand to combine. Drop batter by tablespoons into hot oil and brown on both sides. Serve hot or cold.

Preparing piñons

Piñon nuts have a tough shell (they may need to be soaked overnight before cracking). When the shell is removed, (crack with a small nutcracker), the nuts can be toasted and salted, or used in candy or pastry. They add their own special flavor to cakes and cookies.

Pistachios

Botanical name: *Pistacia vera*

The pistachio tree, a member of the cashew family, can be grown in home orchards in cooler portions of the Sonoran Desert and in the southeastern corner of the state. There is some commercial development in these areas that backers feel is promising.

A native of Central Asia, the tree flourishes in the Mediterranean area. Iran and Turkey are principal exporters of the nuts.

This hardy, subtropical tree produces a fruit about an inch long that encloses an oval, yellow-green seed called a "pistachio nut." The nuts are pleasing in texture and flavor and are often ground and used in confections, and they are great to munch on. (Most so-called pistachio ice cream made commercially, however, is misnamed. It is imitation pistachio ice cream and is made with almonds.) Pistachios are sometimes marketed after being tinted red or coated in salt.

For growing at home, the recommended variety is Kerman.

Pistachio ice cream

This old-fashioned ice cream recipe is a delicious holiday idea when the dish is tinted with a little green food coloring.

¼ pound pistachio nuts
1 cup sugar
4¼ cups cream
1 teaspoon almond extract
Green food coloring
⅛ teaspoon salt

Shell nuts and blanch them by pouring boiling water over them, draining and rinsing. Purée and combine with ¼ cup sugar, ¼ cup cream and a few drops of food coloring. Stir to dissolve sugar. Heat 1 cup cream and stir in ¾ cup sugar and the salt. Stir until dissolved. Cool. Add the pistachio mixture and 3 cups cream. Freeze in 2-quart ice cream freezer. Makes 1¾ quarts.

Sunflower

Sunflowers

Botanical name: *Helianthus annuus*

As many supermarket patrons well know, sunflower seeds are an excellent, nutty-flavored seed to enjoy in place of nuts.

Sunflower plants flourish in desert areas of all elevations when they can get sufficient water. Some varieties, such as Russian Giant and Mammoth, produce flower heads a foot or more in diameter and several pounds of seeds.

Native Americans turned the daisy-shaped petals into a yellow dye and the seeds into black and purple dyes. To obtain the nutritious kernels, the Indians beat the dried heads of the mature flowers with a stick. Almost 50 percent oil, the seeds were then parched and ground into meal.

For modern eating, home-grown sunflower seeds may be toasted under low heat in an oven (salted, if desired) and used as peanuts, but gadgets to shell them do not appear to be available. The best way to get to the kernels is likely with the teeth.

The *Helianthus tuberosus* produces tubers that are called Jerusalem artichokes or sunchokes, which make an excellent root vegetable. However, its flowers are too small to produce edible seeds.

Tepary beans

Botanical name: *Phaseolus acutifolius*

This interesting and delicious native bean, the seed of the tepary plant, has long been a favorite of the Papagos. Smaller than the kidney bean, it comes in various colors.

The name is based on the Papago word for tepary, "pawa," but the bean also has long been enjoyed by Pimas, Yumans, Hopi and other Southwestern Indians. The plant grows wild in canyons, and requires little water when cultivated.

An excellent source of protein, teparies are preserved by drying. Recipes suitable for pinto beans are quite tasty when prepared with teparies, which may be purchased in specialty stores. Crops are ready in early summer, and may be replanted for a second crop before frost.

The beans are high in carbohydrates, and contain a good supply also of potassium, iron and phosphorus. They are about 25 percent protein.

Spiced tepary beans

Tucsonan **Stephanie Daniel** *prepares these desert beans for supper with tomato sauce and spices.*

1 pound teparies (white or brown)
1 24-ounce can tomato sauce
1 large ham bone
2 medium yellow onions, chopped
3 garlic cloves, minced
3 teaspoons oregano
2 teaspoons basil
1 bay leaf
   Salt, pepper

Soak beans overnight in cold water to cover. Drain. Add fresh water, tomato sauce, ham bone, onion and garlic. Season with oregano, basil and bay leaf. Boil beans over medium-high heat for 1 hour. Reduce heat. Cover partially and simmer for 3 to 4 hours, stirring occasionally. Add water if necessary, and salt and pepper to taste.

Walnuts, black

Botanical name: *Juglans major*

The towns of Nogales, both Arizona and Sonora, bear the Spanish name for walnut, and indeed the Arizona black walnut likes the elevation and general climate of that area. The walnuts also are found elsewhere in the state, along streams and in canyons with elevations from 2,000 to 7,000 feet. The trees are widespreading with some growing 50 feet tall. The nuts are ready for gathering in the fall and are very rich in oil. They are smaller than English walnuts and harder to crack, but the flavor is quite good.

It is difficult to transplant the wild trees and grow them at home because of their deep taproots, but they can be grown from seed in a container before transplanting.

Other seeds, desert beans

Buffalo gourd seeds

Botanical name: *Cucurbita foetidissima*

Also known as coyote melon and wild gourd, the baseball-sized buffalo gourd long was prized for its seed, which contains more than 40 percent oil. It is listed here rather than in the melon classification because its seeds are of interest as food.

The Indians used the oil in cooking. The seeds also can be ground, steamed and seasoned for a palatable protein source. The gourds are straw-colored with yellowish stripes. They ripen from September through November. The flesh is skimpy and inedible.

The gourds grow at elevations from 1,000 to 7,000 feet. They require little water and currently are being cultivated for studies at the University of Arizona as a possible source of industrial oil. The roots furnish commercial-quality starch.

Devil's claw seeds

Botanical name: *Proboscidea parviflora*

After the summer rains, the devil's claw springs up in Arizona and other Southwestern states. It is a small plant that produces a fruit or pod resembling a miniature green elephant's trunk. When young and tender, the pod can be steamed or pickled like okra.

Later, when the fruit dries, it turns quite dark and splits to reveal dark or light seeds that are edible and somewhat like large sesame seeds. The seeds are more than a third oil.

Today, the Papagos and Pimas still cultivate the devil's claw for their pods, which they dry and weave into their baskets for a dark accent.

Ironwood beans

Botanical name: *Olneya tesota*

The ironwood tree shuns the cold, and is considered an indicator of a suitable climate for raising citrus. The tree can grow to about 30 feet and provides an attractive accent to urban yards.

The shade from ironwood leaflets is heavy for a desert tree. Small blue-to-violet spring flowers are followed in June by pods containing one or two seeds. These are edible and fairly palatable. (See palo verdes for instructions.)

The name ironwood derives from the fact that ironwood's fine-grained heartwood is very hard. It is excellent for figurine-carving.

Jojoba nuts

Botanical name: *Simmondsia chinensis*

The jojoba is a small native desert shrub with oval, gray-green leaves. Kin to the boxwood, the jojoba is also known as the goat nut or coffee nut. The acornlike jojoba nut grows to about an inch in size and is close to being 50 percent oil.

The jojoba is found at elevations of 1,000 to 5,000 feet. When cultivated, it grows to about six feet tall, and makes an attractive hedge.

The nuts, which appear on the female plants from midsummer to late summer, can be roasted and eaten in that form, or they can be ground and added to baked goods. However, it is recommended that they be used only in small quantities, because of possible diarrhea. The nuts are a little bitter.

Nowadays, jojoba oil is used in cosmetics, and there is considerable research into possible industrial uses for it, especially as a substitute for sperm whale oil.

From the microwave

The microwave oven can be used to great advantage when preparing dishes with fruits and nuts. The same quick-cooking characteristics that retain nutrients in vegetables cooked by microwave apply, and the texture and color will be as appealing as could be desired.

These suggestions compiled by home economist **Kathryn W. Kazaros,** microwave cooking columnist for The Arizona Daily Star, show the scope of fruit and nut cookery for the microwave.

Cornish hens with pecan stuffing

¼ cup chopped onions
⅓ cup thinly sliced celery
3 tablespoons butter or margarine
⅓ cup water
⅓ cup chopped pecans
1 cup herb-seasoned stuffing mix
2 Cornish hens, giblets removed
2 teaspoons butter or margarine
2 teaspoons Kitchen Bouquet sauce

In a 1-quart measure combine onion, celery and butter. Microwave at high for 2½ to 4 minutes or until vegetables are tender. In a 1-cup measure, microwave water on high for 45 seconds to 1½ minutes, or until boiling. Add to vegetable mixture. Stir in nuts and stuffing until evenly moistened. Stuff hens. Place breast-side down on rack on roasting pan or baking dish.

In custard cup melt butter at high for 30 to 45 seconds. Stir in Kitchen Bouquet. Brush hens with half of mixture. Cover with wax paper. Microwave at high for 6 to 8 minutes, rotating dish after half the cooking time. Turn hens over, brush with remaining mixture. Microwave for 6 to 8 minutes, or until juices run clear and legs move easily, rotating dish once. Serves 2 to 4.

Pecans

Almond-ham casserole

7 ounces macaroni, cooked and drained
½ cup milk
1 10¾-ounce can condensed cream of celery soup, undiluted
1 cup dairy sour cream
2 teaspoons prepared mustard
1 green onion, including top, chopped
⅛ teaspoon pepper
2 cups cooked ham, diced
¼ cup slivered almonds
Paprika

Combine all ingredients except almonds in a 2-quart casserole. Microwave for 3 to 5 minutes on high. Stir. Top with almonds and sprinkle with paprika. Return to microwave and cook on high for 4 to 5 minutes or until bubbly. Let stand for 3 to 5 minutes. Serves 6.

Baked fish, lemon-rice stuffing

Fish:

1 2-pound dressed fish
1 teaspoon salt
   Paprika
   Lemon-rice stuffing

Clean, wash and dry fish. Place in large glass baking dish or serving platter. Stuff fish loosely with rice and close opening with toothpicks. Sprinkle lightly with paprika and salt. Place small amount of aluminum foil around tail of fish to keep it from overcooking. Cover dish with plastic wrap. Microwave for 4 minutes. Turn dish and continue cooking for 4 to 5 minutes, or until fish flakes easily when tested with a fork. Allow covered fish to stand for 5 minutes before serving. Serves 4.

Filling:

¾ cup chopped celery
½ cup chopped onion
¼ cup melted fat or oil
1½ cups water
   2 tablespoons grated lemon rind
   1 teaspoon each paprika and salt
      Dash thyme
1½ cups precooked rice
⅓ cup sour cream
¼ cup diced peeled lemons

Place celery, onion and fat in a small glass bowl. Cook in microwave for 3 to 4 minutes or until done. Bring water, lemon rind, salt, paprika and thyme to a boil, cooking for 3 to 5 minutes on high. Add rice and stir to moisten. Cover and let stand for 5 to 10 minutes or until liquid is absorbed. Add vegetable mixture, sour cream and lemon and mix lightly. Stuff fish as directed. Excess stuffing can be placed around the outside of the fish.

Chicken in applesauce

1 3-pound broiler-fryer chicken
1 cup applesauce
   Paprika

Place bird breast-side down on roasting rack. Cover with wax paper. Cook bird on high for 10 minutes. Invert, cover with applesauce and sprinkle with paprika. Insert a microwave thermometer or probe between the inner thigh and body of the bird, away from the bone. Continue cooking on high for 10 to 12 minutes, or until thermometer reaches 170 degrees. Remove from oven and cover with aluminum foil, shiny side toward the food. Let stand for 5 to 10 minutes, or until bird reaches 185 degrees. Salt to taste. Serves 4.

Strawberry pie

1 9-inch baked pie shell
6 cups fresh strawberries (1½ quarts)
1 cup sugar
3 tablespoons cornstarch
½ cup water
   1 3-ounce package cream cheese

Mash berries in food processor or with a fork to measure 1 cup. Stir together sugar, cornstarch, water and crushed strawberries. Cook on high, uncovered, in a 4- to 6-cup measuring device for 5 to 7 minutes or until mixture thickens and boils. Stir once during cooking. Let glaze cool.

Place cream cheese in a small bowl and cook on high for 30 to 50 seconds or until softened. Beat cream cheese until smooth; spread on bottom of pie shell. Fill shell with remaining berries; pour cooled berry glaze over top. Chill at least 3 hours or until set. Serves 6.

Note: Raspberries may be substituted for strawberries.

Toasted almonds

Place 1 cup almonds or other nuts on paper plate and microwave at 50 to 70 percent power for 2 to 2½ minutes, or until hot.

Four-fruit pizza

1 roll refrigerated cookie dough
1 8-ounce package cream cheese
6 tablespoons sugar
1 pint canned apricots
2 tablespoons cornstarch
½ teaspoon pumpkin pie spice
½ cup currant jelly
1 pint fresh strawberries, cut in half
½ pint fresh raspberries
1 canned pineapple slice

Remove metal ends from cookie dough package. Heat wrapped cookie dough on low in microwave oven for 3 to 5 minutes, turning over once. Unwrap and pat into a greased, 12-inch pizza pan. Bake in conventional oven at 375 degrees for 15 to 16 minutes. Cool.

Remove cream cheese from wrapper. In small bowl, heat cream cheese on high for 30 to 50 seconds, or until softened. Stir 4 tablespoons sugar into the cream cheese. Spread on crust. Drain apricots, reserving ¾ cup syrup (use some syrup from canned pineapple if needed).

In 4-cup glass measure combine cornstarch, remaining sugar and spice; stir in reserved syrup. Add jelly. Cook, uncovered, on high for 3 or 4 minutes, until boiling. Stir halfway during the cooking. Arrange fruits on top of cheese; spoon on cooled syrup. Chill for about 3 hours. Cut in wedges. Serves 10.

Apricot crisp

4 cups fresh apricot halves
¼ teaspoon salt
1 cup sugar
½ cup flour
½ cup quick-cooking oats
1 teaspoon cinnamon
⅓ cup butter or margarine
    Sweetened cream or ice cream

Place fruit in ungreased microwave-safe shallow baking dish. Sprinkle with salt. In a separate bowl, measure sugar, flour, oats and cinnamon. Add softened butter and mix thoroughly until crumbly. Sprinkle over fruit. Bake for 8 to 10 minutes on high. Spoon juice from the outer perimeter over the fruit. Serve with ice cream or whipped cream. Serves 4 to 6.

Note: Apple slices may be substituted for the apricots.

Poached pears

4 Arizona Bartlett pears
½ cup port wine
    Nutmeg
    Whipped cream (optional)

Wash and pare the ripened pears, taking care to leave stem on, if possible. Place pears upright in deep individual dessert glasses or wine goblets that are metal-free (see note below). Pour wine over pears and sprinkle with nutmeg. Cover and cook on high for 6 to 8 minutes or until pears are tender. Top with whipped cream and serve to 4. Great hot or cold for breakfast, brunch or dessert.

Micro-hint: Be especially careful not to use good crystal. It is made with lead and that will get so hot in the microwave that it will shatter the glass.

Peaches flambé

¼ cup apricot jam
3 tablespoons sugar
¼ to ½ cup water
4 large peaches, peeled and sliced
1 teaspoon lemon juice
¼ cup brandy
  Vanilla ice cream

Combine jam, sugar and water in microwave-safe casserole or serving dish. Cook on high for about 2 to 5 minutes or until syrupy. Add peaches and lemon juice, continue cooking on high until peaches are tender. Pour brandy in wine glass and heat on high for 30 seconds or until hot. Do not boil. Pour brandy over circumference of the dish. Ignite. Spoon peaches over individual servings of ice cream. Serves 4 to 6.

Cherries jubilee

1 pint jar canned dark sweet cherries
¼ cup rum
  Grated or slivered rind of one orange
1 tablespoon cornstarch
¼ cup brandy, cognac or orange liqueur
  Vanilla ice cream

Drain liquid from cherries, and reserve. Pour rum and orange peel over cherries. Add cornstarch to reserved liquid; stir until no lumps remain. Add cornstarch mixture to cherries. Cook on high for 3 to 5 minutes or until thickened and heated through. Stir well. Place brandy in snifter or crystal glass, heat on high for 30 seconds or until hot. Do not boil. Pour hot brandy over circumference of the dish. Ignite. Serve over vanilla ice cream. Serves 8 to 10.

Baked apples

4 medium apples, cored
¼ teaspoon brown sugar
  Butter or margarine
  Cinnamon

Slice a thin circle of peel from the top of each apple. Arrange apples in a circle in a microwave-safe dish or place each apple in individual custard cups or dessert dishes. Spoon 1 tablespoon brown sugar in each cavity. Place a small piece of butter on each apple and sprinkle with cinnamon. Cover and cook in microwave oven on high for 4 to 6 minutes or until apples are tender. Let apples stand for a few minutes before serving.

Applesauce

6 cups apples, pared, cored
  and coarsely chopped (6 to 7 medium)
2 tablespoons water
1 tablespoon lemon juice
½ to ¾ cup brown sugar, firmly packed
1 teaspoon cinnamon

Combine all ingredients in 1½-quart casserole. Cook covered on high for 4 to 6 minutes, or until apples are tender. Mash or put through sieve. Makes 3 to 4 cups.

Spicy pecans

1 tablespoon butter or margarine, melted
1 teaspoon soy sauce
½ teaspoon paprika
¼ teaspoon ginger
⅛ teaspoon garlic salt
1 cup pecan halves or walnuts

Combine all ingredients, stirring to coat each nut. Spread nuts in single layer in 2-quart utility dish. Cook at 50 to 70 percent power for 4 to 5 minutes or until nuts are light brown. Stir twice during cooking time.

Calamondin lime micro-marmalade

3 cups water
2¼ cups sugar
1 cup seeded, chopped and unpeeled
calamondin limes

Place water, sugar and fruit in an 8-cup glass bowl or measuring device. Cook, uncovered, on 50 percent to 70 percent power or until fruit mixture reaches 220 degrees. If mixture approaches a spillover, reduce the setting to 30 to 40 percent power until the correct temperature is reached. Cooking times may vary from 45 to 75 minutes. Pour marmalade into sterilized jars. Seal with ⅛-inch hot paraffin, or refrigerate jars. Paraffin must be melted conventionally. Makes 2½ cups.

Variation: Use ½ cup calamondin limes and ½ cup kumquats or ½ cup calamondin limes and ½ cup peaches or nectarines.

Micro-hint: A conventional candy or jelly thermometer cannot be used in the microwave during operation. However, the oven can be stopped periodically to determine when the jelly stage has been reached, using the conventional thermometer.

Microwave candy thermometers are currently available. Some ovens also have a separate candymaker probe attachment.

Apple-cranberry sauce

1½ pounds tart cooking apples
8 ounces (2 cups) cranberries
1 to 1¼ cups sugar

Wash, pare, quarter, core and slice apples. Place them in a large glass or ceramic cooking vessel. Look over cranberries, discard soft ones and remove stems. Wash in cold water. Add berries and sugar to apples. Cover and cook on high for approximately 10 to 12 minutes. Remove from oven and let stand for 5 to 10 minutes to thicken. Serve warm or cold. Makes 5 to 6 servings.

Microwave strawberry refrigerator jam

Jackie Wyatt of Sierra Vista shares this method for preparing jam in the microwave. It can be used with other berries as well.

2 cups strawberries, washed and
hulled
1½ cups sugar
2 teasooons powdered fruit pectin

Slice berries in a 2-quart bowl. Mash well. Stir in sugar and pectin. Microwave for 3 to 4 minutes on high or until mixture comes to a full, rolling boil. Reduce setting to 6 (medium) and microwave 5 minutes or until slightly thickened. (It will thicken further as it cools.) Pour in glasses, cover with plastic wrap and store in refrigerator.

Drying fruits

Drying is an ancient method of preservation that has returned to prominence. In Arizona's dry climate it is especially useful.

Drying is easily done out of doors on wooden trays or on screens, with a covering of cheesecloth or nylon netting. Home-built dehydrators, commercially made dehydrators or the kitchen oven also give satisfactory results.

When drying citrus, slice the fruit first. Strawberries also should be sliced. Fruits such as cherries, dates, apricots, grapes and plums — fruits whose skins can be eaten — may be cut in half or sliced and their seeds removed before drying.

Figs, and cherries and plums (if not cut and pitted), as well as grapes left whole, can be dipped in boiling water to crack their skins before drying.

Fruits that darken, such as apricots, apples, pears and peaches, can be pretreated with an ascorbic acid solution, a commercial dip or with sulfur. (See sulfur instructions below.) If using ascorbic acid, 1 teaspoon ascorbic acid crystals per pint of water can be used. For commercial dips, follow instructions on package.

Or, fruits that tend to darken may be syrup-blanched before drying. To do this, bring to a boil 1 cup sugar, 1 cup corn syrup and 3 cups water. Prepare fruits for drying and drop into the syrup. Simmer 10 minutes; remove when cooled somewhat, drain and dry.

Unsulfured fruits that have been dried out of doors should then be heat-treated for 30 minutes in a 150-degree oven, to pasteurize or destroy unwanted organisms.

For oven-drying, the temperature should be 140 to 150 degrees. The trays, which should be about 1½ inches smaller than the inside of the oven for good air flow, should be shifted around every half hour or so, to ensure even drying. The trays also should have 3 inches or more of vertical clearance from each other.

Cool fruits thoroughly before storing in glass jars or other insect-proof containers. If sulfur has been used, put fruit in plastic bags before storing it in metal containers. Containers should be stored in dark, dry, cool place.

How to sulfur fruits

Sulfuring to prevent darkening has both advantages and disadvantages. It cuts down on spoilage. It decreases the loss of vitamins A and C, but increases the loss of thiamin.

Sulfur is available as flowers of sulfur or sublimed sulfur at some pharmacies. It must be used out of doors because the fumes are irritating. In addition, wooden trays are required because sulfur fumes discolor and corrode most metals.

Stack the wooden trays an inch or two apart on wooden blocks or bricks, to allow the fumes to circulate, with the bottom tray about 4 inches above the ground. Cover all with a cardboard box 1½ inches larger than the stack. Cut a slash at the top to allow the fumes to exit, and an opening on the opposite side at the bottom.

Put the required amount of sulfur in a metal container (a 1-pound coffee can or a disposable metal pie plate will do). Set the container beside the lower opening and light it. When the sulfur has burned, close the openings and start counting time.

The recommended amount of sulfur is 1 teaspoon for each pound of prepared fruit. Allow about half an hour for sliced fruit, 1 to 2 hours for half or whole fruits. After the sulfuring time is up, dry the fruit with sun power. (The same trays can be used for drying, if desired.)

Fruit leather

Sometimes called fruit jerky or fruit rolls, fruit leathers are made to be eaten with the hands. Just tear off a piece and enjoy. They are a high-nutrient substitute for candy.

To make fruit leather, purée fruit, adding liquid to make a mixture thin enough to pour. Add sweetening to taste (sugar or honey) and pour about ¼-inch deep onto a cookie sheet lined with plastic.

Dry in the sun, a dehydrator or an oven until leather can be pulled away from paper. Sun-drying takes about a day or more, depending on temperature and humidity. (Be sure to protect fruit from insects and take cookie sheet in at night.) A dehydrator set at 120 degrees takes about half a day; an oven set at 140 takes about 4 or 5 hours.

Leather will peel back easily from plastic when ready to store. Roll up the leather in the plastic, twist the ends and store in refrigerator. Unroll as needed and break off a piece to eat. Or, roll leather as for jelly roll, and overwrap with plastic; cut in crosswise pieces to serve.

If desired, rolls may be stored in plastic bag in cool dry place or overwrapped and frozen.

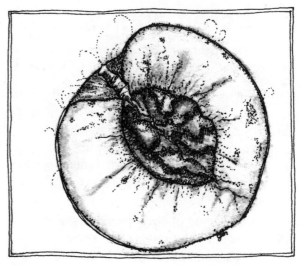
Peach

Tutti-frutti leather

Trudy Lucas *of San Manuel makes her fruit leather with a variety of fruits.*

½ pound fresh plums
¾ pint fresh strawberries, hulled
¾ pound fresh apricots
¾ pound fresh peaches, peeled
  Sugar

Seed plums, apricots and peaches. Slice fruits into separate containers. You will need about 1¼ cups each. Sweeten plums and strawberries with 1 to 2 tablespoons sugar. In separate saucepans, bring each fruit slowly to a boil, stirring constantly to prevent scorching. Remove strawberries from heat but continue cooking peaches, plums and apricots for 3 to 5 minutes, stirring constantly. Purée the fruits in blender. Cool to lukewarm.

Cover two jelly roll pans (15½ by 10½ inches) or cookie sheets with plastic wrap. Pour equal amounts of the four fruit purées onto prepared surfaces, gently swirling together for a pretty, marbled effect. Dry (see directions above).

To store, roll leather in plastic wrap; wrap entire roll again and seal tightly. Makes two 15½-by-10½-inch sheets.

Sugared apples

Stella Hughes of Clifton sends along her method of preparing dried apples to use during the winter — but the results are so good, the apples may not last that long.

Peel as many apples as you can dry at one time. As each apple is peeled put it in a large dishpan with cold water to which ½ cup vinegar has been added (or use Fruit Fresh), to prevent darkening. When apples are peeled, drip-dry on paper toweling before slicing.

Mix 1 pound of dark-brown sugar to each 15 pounds of apples. Next, toss several teaspoonfuls of cinnamon with the apples. This is a little messy and there will be sugary juices left after placing slices on plastic wrap on drying trays. Slant trays at a slight angle to the full sun. The sloping roof or a shed or porch with a southern exposure is a perfect place. (Or even inside a sunny window.) Cover fruit with nylon netting to discourage insects. Dry for several days, taking trays inside at night if there is a chance of moisture. Store fruit in airtight containers.

Cinnamon candy apples: Peel 15 pounds of apples, following sugared apple directions for preventing darkening. Drip-dry on paper toweling and slice. Melt one package of cinnamon hot candies in ¼ cup water and add to apple slices. Dry as outlined for sugared apples.

Lemon apples: Peel 15 pounds of apples, following sugared apple direction for preventing darkening. Drip-dry on paper toweling and slice. Defrost one large can of frozen lemonade and pour it over sliced apples and toss until all slices are coated. Dry as outlined for sugared apples.

Honey pears: Peel 15 pounds of ripe, but not overripe, pears, following sugared apple directions for preventing darkening. Drip-dry on paper toweling before slicing. Pour 2 or 3 cups honey over pears in a dishpan and toss until each slice is coated with honey. Dry as outlined for sugared apples, allowing about a week to dry. Pears are chewy, never crisp.

Strawberry-peach delight

Nancy Pace of Mesa is a strong believer in making fruit leather. Her favorite combination is strawberries with peaches.

Combine overripe strawberries and peaches in a blender in whatever proportion you desire. Line a cookie sheet with plastic wrap and pour mixture on it to a depth of about ¼ inch. Let it air dry in the sunshine with a thin, porous cloth over the top, but not touching the fruit. Roll up with the wrap (after air-drying) and freeze. To eat, break off frozen pieces as desired.

Fruit leather gifts

Former Iowan **Mabel W. Flint** of Duncan learned to make fruit leather when she moved to Arizona three years ago, and now enthusiastically recommends it for gift-giving.

She writes, "I slipped some peach leather into my bag when I flew back in November to visit my 88-year-old father in a retirement nursing home. He and other Midwest relatives' reaction was, 'This is all we want for Christmas gifts from now on!'"

Candied citrus peel

Tucsonan **Virginia Fraps Hodge** *makes candied peel following this method.*

2 cups grapefruit, orange, lime
   or lemon peel, cut in narrow strips
   Water
1 cup sugar

Place strips in heavy pan with 1½ cups cold water. Bring slowly to boiling and simmer for 10 minutes. Drain. Repeat process 4 times, draining well each time.

Make a syrup by bringing ½ cup water and 1 cup sugar to a boil. Add peel and boil until all syrup is absorbed and the peel is transparent, about 45 minutes. Shake peel with granulated sugar in a closed container, then spread on racks to dry. (A good drying device is a big square of window screen, stretched over a shallow box.)

Note: An average-size grapefruit yields about 2 cups strips.

Metric measure

As the country inches (or rather "millimeters") reluctantly toward the metric system, measuring devices for the kitchen are gradually being changed.

For several years, manufacturers such as Corning have been making glass measuring cups with dual measurerments: old-style, standard U.S. cups and fractions of cups on one side and the new metric measurements — liters and fractions of liters on the other. Metric measuring spoons should be available soon.

But there's no need to panic: The changes are slight. A metric cup of 250 milliliters is not much larger than our old-style cup, and the metric measuring spoons are minimally different from our old-style measuring spoons. The liter is not much bigger than a quart. Rarely is it critical in a recipe whether the amount in a cup or teaspoon is level.

In this book, when recipes call for cups, teaspoons, tablespoons etc., use the measuring spoons and cups that you have in the kitchen: old-style or new.

Bibliography

For those who wish additional information about the native and cultivated fruits and nuts of the Sonoran Desert, here is a list of useful and authoritative sources. Reading them has enriched my knowledge of and delight in our desert country.

"American Indian Food and Lore," Carolyn Niethammer, with drawings by Jenean Thomson; Collier Collier Books, 1978.

"Arizona Cook Book," Al and Mildred Fischer, Golden West Publishers, 1968.

"Citrus," Richard Ray and Lance Walheim, with many photographs; HP Books, 1981.

"Citrus Recipes," Al and Mildred Fischer, Golden West Publishers, 1980.

"Desert Plants," quarterly published by the University of Arizona for the Boyce Thompson Southwestern Arboretum.

"Discovering the Desert," William G. McGinnis, University of Arizona Press, 1981.

"Gardening in the Sun," R.B. Streets, Men's Garden Club of Tucson, 1975.

"Hopi Cookery," Juanita Tiger Kavena, University of Arizona Press, 1980.

"New Western Garden Book," Sunset, Lane Publishing Co., 1979.

"Plants for Dry Climates," Mary Rose Duffield & Warren D. Jones, with many photographs; HP Books, 1981.

"Pueblo & Navajo Cookery," Marcia Keegan, with photographs by the author; Earth Books, 1977.

"Seri Indian Food Plants: Desert Subsistence without Agriculture," R.S. Felger and M.B. Moser, Ecology of Food and Nutrition, 1976.

"Sonorensis," periodical from the Arizona-Sonora Desert Museum.

"The Cacti of Arizona," Lyman Benson, The University of Arizona Press, 1969.

"The Sonoran Desert," Roger Dunbier, University of Arizona Press, 1968.

"Trees and Shrubs of the Southwestern Deserts," Lyman Benson amd Robert A. Darrow, revised and expanded; The University of Arizona Press, 1981.

"Western Edible Wild Plants," H.D. Harrington, drawings by Y. Matsumura; University of New Mexico Press, 1972.

— Sandal English

List of contributors

A
Abrams, Linda
Accomazzo, Betty
Allen, Florenia
Alterman, Marsha
Aubrey, Christopher
B
Bach, Virginia E.
Barkley, Lela
Barnette, Kirk
Battiste, Hazel M.
Bergsma, Bev
Bideaux, Jeanette
Birkett, Betty
Bishop, Barbara
Bixler, Marilla
Black Janet
Blattner, Eleanor Beaulac
Block, Josephine
Bonney, Louise
Boyer, Mary
Bradley, Rosemary
Brammer, Mary Lou
Brice, Connie
Brinkerhoff, Ruth
Brookbank, George
Brooks, Henrietta
Brown, Jack
Brunton, Ruth C.
Burden, Sophie
Burgess, Martha "Muffin"
Burks, Mary Love
C
Carlton, Glenn N.
Cheung, Susanne S.
Christensen, Karen
Clark, Amalia Ruiz
Clark, Lenora
Clayton, Syd
Cluff, Nettie
Coatsworth, Hazel
Coca, Donna
Cole, Lunette D.
Condon, Mary Ellen
Cosor, Sue
Coulter, Mary A.
Cowley, Betty
Crane, Betty
Creason, Nada R.
Criley, Mae
Curiel, Gwen
Curley, Jessie
D
Dameron, Dianna
Daniel, Stephanie
Daniels, Agnes
Daniels, Leslie
Dean, Roberta
DeBell, Robin
Decker, Laurel Collier

Della Betta, Leo
de Long, Shari
de Santiago, Lupe
Diers, Judy
Dolph, Roberta
Dotson, Helen
Drees, Mahina
Drorbaugh, Elaine
E
Earl, Winogene W.
Eichinger, Rae
English, Tres
Entrekin, Norma
Eppinga, Jane
Eppinga, Lucia
Estes, Beth
F
Faris, Betty J.
Ferguson, Elizabeth
Fienhold, Alberta
Filer, Ella
Fischer, Jeanne A.
Flint, Mabel W.
Foster, Berenice
Franklin, Agnes
Fraser, Cyndi
Furlong, Nicholas and Charlotte
G
Gallagher, Melba M.
Gawin, Daphine G.
Gekas, Mary
Gibbs, June C.
Gillin, Rose and Carl
Gladden, Beth Oldfather
Goode, Lisa
Grebner, Lance
Green, Dotty M.
Grizzle, Mamie
H
Haller, Janie
Halper, Joan
Hardy, Mary A.
Haverstick, Maxine
Hefty, Marjorie
Henry, Bette
Hill, Doris C.
Hinsdale, Rosejean
Hodge, Virginia Fraps
Hoefle, Selma
Hollis, Muriel E.
Horner, Florence E.
Houpis, Anne
Hughes, Margarette
Hughes, Stella
Hunter, Jere
Hulse, Leonie
Hurcombe, Esther
I
Ingram, Helen

Irwin, Hester
J
Jackson, Dorothea B.
Jancic, Inge
Johnson, Eunice I.
Johnson, Marie
Johnson, Val
Jones, Audrey
Jones, Eileen
Jones, JoAnne
Jones, Mary C.
Jones, Nan
K
Kata, Jacquie
Kazaros, Kathryn W.
Kececioglu, Lorene
Kelly, Katie
Kendt, Nancy
Kightlinger, Ellen M.
Kilmer, Dorothy E.
Kipps, Julie
Knuth, Susan A.
L
Larsen, Helen
Larson, Sabina
Lemen, Idola
Lilleboe, Carmen
Lopez, Anne
Losee, Renee
Loso, Mary
Lucas, Trudy
Lundberg, Hanna
Lynde, Geraldine
M
MacMillan, Larry
Mansker, Kathryn
Marsh, Carline
Martin, Iola
Martin, Susan
Mattausch, Eleanor
Mattausch, Etta
Mauntel, Billie
Maycher, Rosario
Mays, Priscilla J.
McGill, Julee
McLennan, Juanita
Melgren, Geneva
Metzger, Mary
Migliacci, Sandy
Mikel, Terry
Mitchell, Nina
Molina, Herlinda
Mollison, Anna Marie
Moreno, Lorraine
Musnicki, Vicky
N
Nagel, Carlos
Niethammer, Carolyn
Noon, Carol
Nyburg, Mary

177

Prickly pear fruit

Recipe index

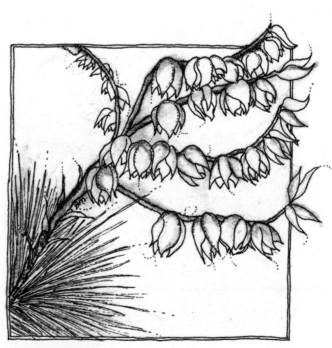

Yucca blossoms